Nucleus

NUCLEUS

Reconnecting
Science and Religion
in the
Nuclear Age

by Scott Thomas Eastham
Foreword by Raimundo Panikkar

BEAR & COMPANY
SANTA FE, NEW MEXICO

Copyright © 1987 by Scott Thomas Eastham
Foreword Copyright © 1987 by Raimundo Panikkar

Library of Congress Cataloging-in-Publication Data

Eastham, Scott, 1949-
    Nucleus: reconnecting science and religion in the nuclear age.

    Bibliography: p.
    1. Nuclear warfare—Religious aspects—Christianity.
I. Title.
BR115.A85E19 1986   261.8'73   86-22265
ISBN 0-939680-31-9

Printed in the United States of America

BEAR & COMPANY
P.O. Drawer 2860
Santa Fe, New Mexico 87504

For my Mary, and
our child, as yet
unborn.

Table of Contents

Nucleus

Foreword

The Challenge of Religious Studies To the Issues of Our Times

by R. Panikkar

Scott Eastham's *Nucleus* is an intelligent inquiry into the religious aspect of one of the most pressing issues ever to confront humankind: that of its own extinction. The book offers a new point of departure and a welcome agenda for collegiality. By focusing his approach to the nuclear issue on the full scope of human relationships, Eastham also highlights a crisis facing religious studies as an academic discipline—and, indeed, facing the university itself as an institution originally intended for the unfettered pursuit of truth. It is to this problematic that I shall devote my introductory remarks.

The Crisis:
An Inadequate Perception of the Situation

My thesis here is a simple one. The crisis today undermining the identity of religious studies in the university setting comes from the high demands placed upon this emerging discipline. We ask religious studies not only to direct its efforts toward solving ancient riddles, but also to come to grips with the real challenges today confronting human and planetary life. Not so very long ago, programs and departments in religious studies sprang up all over the academic world in Europe and North America, raising high expectations on many fronts. Today this very discipline seems to be in danger of becoming an accepted but minor academic sideline for those few universities which can still afford the "luxury" of truly humanistic studies. This much said, let me briefly outline what I take to be the principal challenges today facing the study of religion.

The proper field of religious studies, as I see it, is *the religious dimension of Man*.* To be sure, this religious dimension transcends the enclosures of the academic world and of institutionalized religions, but it is often expressed in these institutions. Yet living institutions do not thrive by

---

*Once and for all I state that for me the word Man means the androgynous human being and not the male element which has hitherto monopolized it—although when using the pronouns, I follow common usage, waiting for the *utrum*, the new gender which encompasses both masculine and feminine without reducing either to a non-human *neuter*. It is not that the masculine stands for the whole Man, but that the whole Man has allowed this domination by the male. The solution is not juxtaposition (he/she, etc.), but integration. For this reason I shall use Man when referring to the *anthropos* as an irreducible reality standing side by side—with all the necessary ontological distinctions—with God and world.

severing their connections and interactions with human life at large. We have before us instances from the natural sciences, which operate in close symbiosis with industry, defense, politics, economics, and many other perceived social needs. Religious studies has no need to legitimize itself by appealing to examples from other disciplines, but by the same token it must be conceded that this discipline runs a considerable risk of losing touch with reality altogether unless it strengthens some similar connection with the genuine human concerns of our time. The study of religion is neither mere scholasticism nor archaeology. Religious studies should feel neither so secure nor so insecure that it neglects interaction with the world at large. Nor, obviously, should it still be clinging to remnants of the aloof superiority sometimes implied by the words *holy, sacred,* and the like, segregating its ultimate concerns from immediate human and secular crises.

In a secular setting, I would describe the religious dimension as the *human face* implicit in any human venture. Mathematics, physics, politics, music—all are activities of the human spirit dealing with particular aspects of reality under determinate perspectives. When they touch, and insofar as they touch what I call the human face, they touch the religious dimension. How these disciplines enhance, help, or transform human life—positively or negatively—is precisely the religious dimension of the phenomenon.

In a more traditional setting, I could describe the religious dimension as the *reference to transcendence* implicit in any human venture. Geometry, chemistry, sociology, architecture— all are activities of the human being focusing on particular objects. When they encounter an elusive element, a *plus* or a *more* irreducible to the proper parameters of their disciplines and yet inextricably connected to them, they meet what we call transcendence, they meet the religious dimension.

I submit that we have here two languages that do not say the "same" thing, but refer to the same problem: the question of the religious dimension of Man. I should simply mention here that the sacred is opposed to the profane, but not necessarily to the secular.

The challenge facing religious studies is to enter the arena of today's life and not shirk its responsibility to criticize and even to inspire actions, people, and issues in the public domain. I should not be misunderstood. I am not proposing to politicize religious studies. I am, however, suggesting that we recover for religious studies the traditional meanings of both *religion* and *studium*, and I am urging that we do not content ourselves with being curators of antiquities or hoarders of ancient treasures. There is something amiss in a society where the ladies keep their diamonds and emeralds in a safe and wear glass baubles in public. And, likewise, there is surely something rotten in Denmark when scholars of religion are conspicuously silent on many of the most important religious issues of our times. There is, undoubtedly, a political facet to religious studies.

I should stress that I am not defending a shallow involvement in discussions à la mode, as if the religious scholar's only task were to offer quick technical fixes or even brilliant theories having no viable roots in history or tradition. To the contrary, we must take up the task of dialogue with the most pressing issues of our times, a task which will require bringing to bear all the precious and indispensable human wisdom accumulated down the generations and throughout the world. Certainly we should not cease poring over the treasures of the past, or developing purely theoretical or disinterested investigations. But we should also be carrying these treasures all the way into the present human world, and exhibiting these jewels in their full and proper setting in the overall enterprise of religious studies.

The challenge is to pass from the past to the present, that is, to undergo the transition from describing what religion has been to understanding what the religious dimension of Man is now, and how it expresses itself in and for our contemporary situation. We need not only to analyze ancient worldviews and diverse mentalities, but also to assimilate these alternative ways of being, knowing, doing, and becoming. We should indeed know as much as possible about the Bantu ancestors, and the Vedic sacrifice, and the Eleusinian mysteries, but we should never forget that all this knowledge is part of the knowledge of ourselves, which has to be integrated not only into our understanding of how homo sapiens *was*, but also how we *are*—not only as individuals but as a collectivity and as a cosmic being. The challenge lies in the fact that our theories about religion and religious issues are not independent from our praxis. Religious studies is not a normal academic discipline. It is a *sui generis* human activity, due precisely to the nature of its subject matter: religion. Let us recall the world scene today in order to see how the arena of our times has basically changed, and how this change now demands a parallel transformation of religion.

The Great Confrontation:
Neither God, nor Nature nor Man, but a Factitious World

The great confrontation of our day is no longer the old struggle with the Divine and/or the human efforts to tame nature. Until recently, God and nature were the great challenges. When confronted by ultimate questions, Man had to wrestle with God, he had to appease the Gods, to entreat them, to obey the divine rules and follow the supreme laws. This was the classical domain of Religion. Through a

kairological shift (which need not follow in strictly
chronological order), when confronting ultimate issues Man
found that nature also had to be known, that her rules had to
be discovered, that humans had to be able to predict her
behavior and observe her laws—precisely in order to bring
nature into a higher form of harmony with the human being.
This is the classical domain of Science. At the same time,
thirdly, people have always been aware of dangers posed by
their fellow human beings, and have had to learn to cope
with one another: women with men and vice versa, one
generation with the next, commoners with their chiefs, and
the like. One "tribe" could always be a menace to another,
whether such tribes were called families, clans, nations,
empires, corporations, or superpowers. This is the classical
domain of Politics. The divine, the cosmic, and the human
constituted the three worlds: the natural habitat of Man.
Religion, Science, and Politics were the three main spheres of
human life. Now, something new emerges. Some human beings
have always threatened others, but Man was not the foe
of humankind, nor was he powerful enough to sweep away
the entire human species or to alter the gene pools of whole
races. When self-understanding ceases to be an authentic
understanding of the Self, the old paradox of knowledge as
the science of good and evil recovers all its diabolical force.

Today, prepared by modern technology (which is more
than just applied science), and triggered by the split of the
unsplittable (the *a-tomos*), the great challenge to Man is an
artifical world which has become independent of its engineer.
The great confrontation is no longer that of Man facing God
or nature or other people, but Man facing the historical,
technological, and scientific forces that a certain human
calculating power has set into motion as a fabricated world.
Before this factitious System, the individual feels more lost

and forlorn than humans ever did before the divine, the natural, or the human worlds. Nobody seems accountable for today's menacing disasters. The System, unlike God(s) or nature or human society, resists anthropomorphizing: it cannot be personalized; it is anonymous; nobody seems to have any real control over it. In this "fourth world" the Gods are silent, the cosmos is dead or dying, and people are convinced of their own impotence. It is pathetic to hear the Reagans, Gorbachevs, Thatchers, and Kolbs of the world saying they would like to have a future free of nuclear weapons, but that they cannot: it is an idealistic pipe dream. What is this power which is uncontrollable by its controllers? We should in this regard by no means confine our thinking to an atomic holocaust; we should bear equally in mind certain disturbing initiatives lately undertaken in psychology, genetics, and biochemistry, which also have their own autonomy and also seem to have escaped human control.

We should, moreover, become increasingly aware of the prisons we have made for ourselves under the guise of protection or security. The very magnitude of modern cities puts up higher walls of segregation than did the medieval gates, especially for those—the majority—who cannot afford to escape to the suburbs. Working under electric lights and artificial climates, working within the megamachine of the industrial metropolis, submitting to the entire web of routines and regulations regarding each and every human transaction makes us terribly vulnerable to any failure of the System and robs us of our human spontaneity in a measure almost unimaginable to those already born into the strictures of this "fourth world." No wonder that, having built such a splendid prison for themselves, people are so easily exasperated when the system of controls breaks down, as it has lately in tourism and space travel.

In short, the total destiny of the human being today is no longer subject mainly to the "will" of God, the "whims" of nature or the "programs" of the politicians, but to the "sphinx" of the dehumanized and artificial technocratic complex. God, nature, and society may still play important roles for many, but for the elites of the technocratic complex their predominance has been precluded. Famine, for instance, which has never been so widespread as in our own day, is now seen neither as a divine punishment, nor as a disorder in the seasons of nature, nor even mainly as a political failure, but rather as a technical and economic problem of logistics.

A Basic Religious Issue:
Civilization as Human Project

It is obvious that the world today is in dire straits. We live in the midst of a global malaise that cries out for us to investigate its hidden roots. The so-called System of modern civilization— the interlocking complex of economic, political, and technological mechanisms—has today come to the point of actually threatening, not only those considered its enemies, but the majority of the peoples of the Earth and, indeed, the life of the planet itself. We are becoming aware that it is more than just a mistake or computational error that has crept in somewhere along the line. The modern predicament alerts us that there is something intrinsically wrong at the very basis of our civilization. As this book makes abundantly clear, the "nuclear issue" is just the tip of an iceberg, the bulk of which is still obscured beneath all the current technical, political, and economic upheavals. At bottom, I submit, the culprit is the prevalent project of civilization itself over the past 6,000 years.

The roots of this problem grip down all the way into the neolithic passage from decentralized agricultural villages to the organization of centralized cities and states, and probably also from the matriarchate to the patriarchate. In so many words, the roots of the problem are tangled up in the urban revolution we have come to call civilization (where the *civitas* becomes the center of "civilized" life). There are certain fundamental differences between *culture* and *civilization*, that is, between village life and city life. While the town needs fences, as its etymology indicates, the village has none, or at the most has gates. While the city is largely parasitic on the agricultural provinces, the village and its people live in symbiosis with the Earth and her rhythms. While every *polis* requires slaves under one name or another, the village requires only a certain type of caste system (which, obviously, can also degenerate). While the city is not self-supporting, the village usually gets by on its own with a natural commerce or exchange of goods. While the modern city demands technology, the village needs only technique. (I make a fundamental distinction between *techné*—technique—which is a cultural invariant, and modern technology, which is a product of modern civilization.) While the city requires political or military or police power to bring it under control, the agricultural tribe or hunter-gatherer cultures make do with hierarchical structures which confer authority. It is only with the city that the schism between authority and power becomes acute. Power implies the capacity for physical, psychological, and epistemological control. The very word authority suggests a kind of sanction that allows things and people to grow and evolve. One has power; one is conferred authority. Yet the village also has its own flaws, and there is no longer any question of totally de-urbanizing civilization or of idealizing village life. What we do need is the paganization of culture, that is, the integration of the *pagi* (the villages, and thus those who dwell there, the "pagans") into the human project of life—which today must overcome not only the dichotomies spawned by overspecialization (including the body/soul split), but also the Man/nature and God/world dualisms. This can be understood and acted upon only if we integrate into our vision of reality perspectives different from the predominant modern one.

Let me put forward just one example. We are accustomed to viewing history as a succession of wars followed by more or less stable truces. We are told that wars have occasioned great discoveries, given opportunities for the exercise of heroic virtues, and provided outlets by which the accumulated aggressive energies of a people were discharged. We were able to live with such rationalizations only because they are partial truths. Down through the ages, the military castes in various cultures have had the sacred duty of keeping the world from falling apart, of saving the *kosmos* from disintegrating into *chaos*. We were told that wars were natural phenomena; or rather, we chose to believe they were, without facing up to the fact that they are civilizational and highly institutionalized phenomena. No other animal species wages war.

Today we are beginning to see the mutation which has taken place in the very notion of war. Since the splitting of the atom, all of these traditional attitudes have begun to undergo radical change. We are finally surmising that the human project must find a new direction. Most of us may not care to, probably should not, and certainly cannot turn back to a simple agricultural lifestyle, but we are equally aware that merely patchwork reforms of the present-day technocratic System will not keep it afloat much longer. What is really emerging here is a mutation in human life altogether, the end of the hegemony of historical consciousness. Unless we undergo this transformation, this *metanoia*, humankind will indeed soon commit terracide. These kinds of questions lie at the basis of this book, and it is the collaborative search for answers to such questions which will eventually help us discover the new direction(s) that we human beings need to survive this nuclear age.

Significantly enough, poets, thinkers, artists, and even religious authorities have for well over a century now been descrying the end of this western civilization. They have tended to have minimal impact on the world situation, mainly because what they foresaw was not yet conspicuous enough to ordinary people. Today a neo-millennial climate is being felt all over the globe, although it differs in many respects and carries with it fewer eschatological elements than former millennial movements. We should be equally wary of both prophets of doom and messiahs of optimism. I am suggesting that we are approaching the end of historical civilization, and at the same time witnessing the discreet dawning of a possible metahistorical or transhistorical culture.

I am making these rather disturbing overstatements (which I have tried to substantiate and qualify elsewhere) in order to show that the problem is basic. It calls into question the very foundations of human civilization, and it requires a cultural metamorphosis, the *metanoia* alluded to above, even to be adequately perceived. At any rate, this is a question of life and death, of human fulfillment or annihilation, of our survival as a species. As such, it is a specifically religious question. It cannot be solved technically, or politically, or economically. It requires a cultural transformation of the first magnitude. This issue strikes to the very basis of human existence, indeed, to the heart of life itself. And if such an issue is not a religious incumbency, I am left wondering what religion is all about.

Definition of Religion:
Way to Peace

There is perhaps a semantic problem here. Should we
confine what we call Religion exclusively to the first struggle,
that of Man with the Divine, and reserve the name of Science
for the second, namely that of Man with nature? Should we
call Politics the art of the dealings of people with one another,
as the third main human activity? What about the fourth and
present world? Is this then to be the field of human and
physical Engineering? Are all four not intermingled? Was
religion not always concerned with the vital issues of a
particular tradition? Are we to limit ourselves to being priests
in the service of the past and shrink from becoming prophets—
or even martyrs—in the service of the future? Can we not
plant our feet firmly in the soil of tradition, fix our eyes on the
horizon of the future, and also begin to raise our voices
amidst the hue and cry of the present? "Religion and
Science" and "Religion and Politics" are classical problems.
"Religion and Engineering," in the sense just indicated, is an
urgent modern issue.

My longstanding definition of religion, which has at least
the advantage of being the shortest, is this: ultimate way. In
Greek, it is a single word: *eschatodos*. If we want to give this
definition a specific content, we can translate "ultimate way"
by "way to salvation"—a respectable traditional word that
stands for liberation, fulfillment, paradise, joy, God, justice,
or any of the other homeomorphic equivalents from the
various religious traditions. The task of religious studies is to
critically investigate these ways, with all their concommitant
sets of symbols, myths, beliefs, rituals, doctrines, and
institutions, which people believe convey the ultimate
meaning of the human pilgrimage or of existence in general.
I am proposing to render the same idea contemporary by

putting it as follows: *religion is the way to peace*—so that here peace becomes the present-day homeomorphic equivalent for all those earlier interpretations of "salvation." This peace is obviously not an exclusively political peace nor an internal concord only. It is a complex and polysemic symbol standing for the cosmic and personal harmony of (and within) reality. It is something like the Pauline "recapitulation of all things in Christ," who is coherently called "peace" in the Epistle to the Ephesians.

Peace is the blend of harmony, freedom, and justice. Some cultures stress freedom, and often only individual freedom. They are sensitive to the constraints that hamper human blossoming from without. Human life is seen as development. The modern ideology of development, which leads us to speak of developed and underdeveloped countries (or worse: LDCs, "lesser-developed countries") is an example.

Some other cultures stress justice, often viewed collectively. They are sensitive to the structural obstacles to the construction of a just social order. Human life is seen as well-being—and this well-being is usually framed only in terms of the material needs of the people.

Some other cultures stress harmony, viewed most often in connection with nature. They are sensitive to the kind of human happiness which comes from playing one's part in the unfolding drama of the universe. Life is the main category, and human integration into that Life the main value.

Freedom points to truth. Justice points to goodness. Harmony points to beauty.

A genuine peace would have to blend these three values, visions, experiences, both in a time-less/time-full (tempiternal) way and in a personal/cosmic manner. The traditional word for this blend is Love. And all of these, I submit, are religious categories.

One of the most urgent tasks for religious studies today is to rethink its proper subject matter. Religious studies simply cannot leave the ultimate queries of humankind to the voluble responses of merely technical solutions, or to analyses generated solely by sciences working within a single worldview—even if it is the prevailing worldview—or to the ideologies of politics, even those of institutionalized religious bodies. This does not mean trying to supplant the technical and scientific efforts, or abandoning all churches altogether. It does however mean that we must become more acutely aware of the present human predicament and its continuity or discontinuity with the world's religious and cultural traditions. The worst service to tradition is to freeze it.

Let me reformulate all this: *the object of religion is not God, it is the destiny of Man*—and of Man not just as individual, but also as society, as species and genus, as microcosm, as one constitutive element of reality which both mirrors and shapes that reality. Life on Earth may not be the ultimate destiny of Man according to some traditions, but even *vita eterna, nirvāna,* and *brahman* depend for their very formulation on the continued existence of this earthly life. If the planet is threatened or obliterated this may not be an ultimate tragedy for the galaxy at large, but it is without doubt a universal religious concern here on Earth. God, a traditional name for the object of religion, may indeed represent the true destiny of the human being, but surely is not the only such symbol. The study of religion properly deals not with any one symbol alone, but with any symbol standing for the ultimate meaning of life and the means to reach it. In a word, the concern of religion is human destiny, and it is this human destiny which is at stake today.

External Challenges

Any contemporary approach to a truly human problem which is not tackled in a cross-cultural way smacks of cultural ethnocentrism. This ethnocentrism is the intellectual heir to the colonialistic attitude of times allegedly past. We need to study the problem of Man, drawing into the study the various self-understandings of the different peoples of the Earth, for the very essence of Man includes precisely this self-understanding. Man is that entity which has self-understanding as its most relevant characteristic. We cannot study Man without integrating the way in which the human being understands itself. Now this self-understanding may vary for different times and cultures. Religious studies offers hospitality, as it were, to representatives of the most diverse worldviews. It allows them to have their say in their respective languages, but it must also recognize that inviting them to come together in this way is a risky venture, one that brings into sharp focus all the questions we have alluded to above. Perhaps here we should pause to take our bearings and sort out the challenges religious studies is facing, as well as the challenges religious studies itself poses to the status quo. Scott Eastham's study is an example of this multiple challenge.

Challenges to Religious Studies:
New Tasks for Religion

If the foregoing reflections are not altogether off target, then the challenges to religious studies as we enter the Aquarian Age may indeed be many, complex, and bewildering. For clarity's sake, I shall group most of them under three headings: as challenges from the world at large, the organized religions, and the academic world.

The world at large obviously confronts religious studies

with a set of new and pressing problems. Peace, justice,
hunger, war, armaments, nationalisms, intolerance, mutual
ignorance, and ideologies of all sorts (political, economic, and
religious) represent a host of problems which cannot even be
approached adequately, let alone understood, unless the
religious dimension is included in the problematic. Human
problems, unlike purely objective problems, cannot be solved
by the Cartesian rule of dividing them into as many minor
parts as possible. The whole is a good deal more than the
sum of its parts. All of the problems just mentioned challenge
religious studies to study them, to contribute to their
clarification, and eventually to take a stance. By and large,
religious studies is not yet sufficiently prepared to tackle
these issues. I see here a theological remnant—much like that
of a church seeking to hold itself above merely human
disputes because it deals with the supernatural destiny of the
people. But those who fear to risk taking sides might already
have taken the wrong one, and are probably going to be
shoved aside. Religious studies cannot be silent on such
issues.

The organized religions are encountering any number of
new situations which oblige them not only to adapt, but also
to transform their very conceptions of themselves and their
roles in the world. Present-day Islam, Hinduism, or
contemporary Catholicism, for instance, cannot be studied
merely from books and traditional or canonical documents.
The internal dynamism of each religion as it undergoes
change demands that religious studies apply new categories
of understanding which will have to be forged in the study of
these new phenomena. I am not so much referring to the
so-called new religions as I am to the new religious spirit
pervading sacred and secular institutions alike. I submit that
these questions are not marginal to religious studies and that,

necessary as it also is, merely sociological research into such phenomena is not enough.

The academic world, finally, puts to religious studies the most embarrassing questions. And within a university, certainly, all the other departments are fully within their rights to do so. Too often, mutual critique is not even attempted because the experts in religion are in many cases ignorant even of the languages of the other disciplines, let alone of the worldviews implied in them. Biology, for example, today presents us with a series of crucial questions regarding the ultimate makeup of the biosphere and the meaning of life therein. Such questions originate in a strictly biological approach to what are apparently purely biological issues, and yet in the long run transcend the realm of those sciences and spill over into every domain of human life. Without more real give and take between disciplines, the study of religion will soon find itself swallowed up into a branch of history, or anthropology, or philosophy, and lose its raison d'être in the academic world. I am not claiming that religious studies has the answers. I am saying that it ought to be helping to frame the proper questions. It is a great challenge. Yet even this is not the whole picture.

Challenges from Religious Studies:
Upsets the Status Quo

Insofar as religious studies is not sufficiently prepared to meet the challenges of the outer world, it has to ready itself for this momentous task. But insofar as it has begun to face these issues—and *Nucleus* is a case in point—then religious studies itself presents certain unsettling challenges to the world at large, to the established religions and to the academic world.

The fact that religious studies is a secular activity, albeit

an academic one, puts it on much the same level as any other
secular affair. It has to deal with human problems in the
fields of industry, politics, government, business, the
judiciary, and so on, without recourse to any higher external
authority. But by the same token, religious studies cannot
artificially truncate itself and enclose itself in a separate
compartment dealing with problems exclusively its own. The
field of religious studies cannot be objectively delimited from
the field of humanity at large, although it primarily concentrates
on a specific aspect of the human enterprise. But this aspect
is just that: an aspect, a perspective, which cannot be
artificially segregated from anything human. Human beings
have a constitutive religious dimension. Every cultural
activity contains more or less conspicuous religious elements
insofar as it is an activity of the whole human being.
Religious studies has something to say regarding atomic
research, consumerism, democracy, economic standards of
well-being, nationalism, justice, and peace, so that this
discipline may often appear to challenge the predominant
ideologies. And if religious studies ceases to do this, it ceases
to perform one of its primary tasks. Even if one were to say
that religious studies is not directly concerned with what
Man *is*, but with what humans *believe*, beliefs qua beliefs (and
not as mere doctrines) cannot be severed from Man, the
believer.

Religious studies similarly challenges religious
institutions, both in their day-to-day operations and in their
guiding assumptions and presuppositions. In variegated
ways and forms, religious organizations are dedicated to the
pursuit or practice of a particular religious ideal. In many
countries these institutions suffer the intervention of the
state, and in many others they have to act within a certain
framework erected by the state. Nonetheless, they try to keep

their autonomy. But scholars of religious studies are in a peculiar position. They are neither mere outsiders nor insiders only. Many of the same problems that occupy the time and attention of the various religious organizations are also taken up by religious studies, but from a different perspective. Religious studies is neither confessional nor normative; it is not apologetics, nor is it even concerned with the apologetics of religion. It is not an assumption of religious studies that religion is necessarily a good thing. It could equally be diabolical or alienating. The thoroughgoing critique of religion belongs essentially to religious studies, and this may often be sensed—rightly or wrongly—as a threat to more than one religious organization. I am not saying that organized religions are inherently exclusivistic and proselytizing, but there is no doubt that the academic discipline of religious studies, particularly within religious organizations, often comes to the table as a rather embarrassing guest.

To the world of academia the challenge sometimes appears so great that it may be perceived as an outright threat. Religious studies as I am describing it does in fact challenge a certain liberal and individualistic scheme of things which has lately become more and more prevalent in the worldwide academic community. Religious studies seems to impinge on the almost totally independent sovereignty of many a faculty or department, not only by pointing out inescapable ethical dilemmas inherent to the various disciplines, but also by uncovering the cosmological, anthropological, and even religious assumptions of the particular sciences. Religious studies asks the disturbing questions which all too many of the sciences, both "natural" and "human," prefer not to hear. Asking whether we should perform experiments on living beings may be a recognized

ethical question, but asking whether atomic fission alters the
natural rhythms of matter may be a cosmological query
issuing from another *Weltanschauung* altogether, and yet is
equally legitimate. In this sense, religious studies tends to
bring the sciences together but it also creates problems for
them which would not arise were they to continue their
investigations in isolation. Or should we let the atomic
scientists do their jobs without interference, even if they end
up by exploding the planet? Human abortion is certainly a
religious question, but the cosmic abortion represented by
fissioning the very womb of the atom is equally a religious
inquiry, and probably a greater one. In short, religious
studies contributes to the unity of knowledge by confronting
the disunity of present-day academic disciplines. To help
overcome the fragmentation of knowledge is one of the major
tasks of our culture.

The Internal Challenge:
The Question of Truth

I have already suggested that the understanding of
religion can no longer be restricted to phenomena of the past,
and that we need to embrace in that same word wholly new
human reactions to the ultimate questions facing human life
today. In my opinion, "History of Religions" is already an
obsolete name for the contemporary study of religion.
Similarly, it may now be incumbent upon religious studies
to recall that *studium* once meant and could mean again the
complete, diligent, and enthusiastic application of our entire
being. Some of us may still remember those memorable
words of Cicero defining *studium* as that *animi assidua et
vehemens ad aliquam rem applicata magna cum voluntate occupatio*:
the intense and enthusiastic activity of the spirit directed to
something with great will, that is, with all the power of our

striving and commitment.

In other words, religious studies as a discipline should be able to overcome the deadening trend of our times and dare to be religious studies, that is, the intellectual and existential study of the roots of our being, our place in the cosmos, and our destiny as human beings implicated in the very warp and weft of the entire reality.

The final challenge, I submit, comes from one inescapable feature of religious studies, a question which can often easily be avoided in other disciplines, but which is actually formative and constitutive of religious studies altogether. It is the axiological aspect. Religious studies can afford to be value free only when being value free is a positive and not a negative value. In fact, religious studies is never value free. It cannot skirt the axiological component. The venture toward an authentic understanding of religious problems cannot stop short of evaluating whatever hypotheses are studied. To be sure, religious studies is not identical to religion; it is not confessional, and yet it is professional. The study of religion can never be quite like the study of crystallography, for instance. This is a delicate question. I submit that there is a *via media* between being normative and being purely descriptive. Religious studies cannot be normative. It has neither the tools nor the power to set the rules. But it is not merely descriptive either. Human genetic manipulation, totalitarian ideologies, social genocide, the defoliation of nature, unjust laws, degradation of human dignity, and many another issue could be adduced in regard to which religious studies simply has to take a stand—a stance that follows directly upon its internal clarification of the problem and analysis of its assumptions. Where other disciplines may remain on the level of description and continue to function, religious studies has to take a stand in order to link particular

issues with the overall picture of human fulfillment. Any
stand taken is of course provisional and open to criticism,
dialogue, and refutation; but it is nevertheless a stand. There
is no neutral ground. Failing to take a stand amounts to a tacit
endorsement of the status quo. This final challenge may seem
daunting: nuclear weapons, abortion, state terrorism,
apartheid, militarization, but also liberalism, capitalism,
socialism, nationalism, and nonviolence may serve as
examples here. It is a risk, and a responsibility.

To put it more pointedly, the present-day scientific
disciplines are not directly concerned with truth. Their
primary concerns are to acquire the most accurate possible
descriptions of objectifiable data, to develop pragmatic
knowledge of how things work, and eventually to put that
knowledge to "good" use. But truth is more than (mathematical)
exactitude, (physical) consistency, (historical) coherence,
(political) efficacy, or (psychological) plausibility. Human life
cannot avoid the question of truth. Religious studies is in this
sense closer to common sense and to human life as it is lived:
the question of truth cannot be brushed aside. And it is
precisely because truth is central and unavoidable to its
concerns that religious studies soon discovers that truth itself
is polyvalent, polysemic, certainly relational, and probably
pluralistic at the ultimate level.

The Nature of Religious Studies:
Cross-cultural and Interdisciplinary

I have been uncapping all these highly debated questions
only in order to bring home the role of religious studies.
There are literally hundreds of excellent books available today
which explore alternatives to the present world order. Many
of these present workable options, and they are a sign of
hope. Such studies not only have thousands of readers, but

indeed reflect the existence of movements and people who are putting the new ideas into practice. These are islands for survival, links to a new and truly human lifestyle. But most are still grafted onto the old System, mainly because it is almost impossible to do otherwise. These new movements too often do not take seriously enough the possible contributions of other cultures, religions, and wordviews. Today we need much more than a handful of isolated options meaningful only to small groups of people reacting to some common frustrating experiences.

It is here that the contribution of religious studies may well be crucial. Religious studies offers the natural/cultural platform where dialogue and possibly mutual fecundation may take place. The different religious traditions of the world simply do not speak the same language (i.e., share the same myth, symbols, context). How can American Indians, for instance, present a case against the construction of a highway or atomic power plant on the grounds that their ancestors were buried on the site and that the proposed development would render communication and even personal identification with them impossible? Mainstream American civilization may yield in one or two places out of political necessity, but the Indians are never going to convince the engineers of the reasonableness of such an "absurd" demand. The retorts are familiar: "Ghosts do not exist," "Land is just a resource, a plot of measurable Newtonian space," "Development means improvement," "We'll pay you for it." That the ancestors may be real or the Earth a living organism is sheer nonsense and superstition for the engineer. Or again, to claim India could solve her protein deficiency by slaughtering her sacred cows amounts, in its context, to suggesting that North America balance its budget by consuming human flesh. These are clashes at the mythic level, involving not just one behavior or another but entire modes of organizing human life. And to suppose that we respect the Hindu, the Christian, or the Atheist without having even an inkling of their values or ideas comes to no more than political expediency.

I propose, therefore, that the discipline of religious studies take on the role of a sort of clearinghouse where contributions from the most divergent disciplines and disparate cultures are brought into dialogue and interaction on their own terms, that is, without undue violence to the respective insights and worldviews of those disciplines and cultures. In so many words, the role of religious studies today is to offer a clearinghouse where the ultimate problems of the human world may be sifted, clarified, discussed, perhaps understood, and eventually solved. The discipline would then be nothing less than the concerted essay to understand Man's ultimate predicament in a world context by all the possible means at our disposal. If religious studies is concerned to study ultimate human values and disvalues, without excluding those of contemporary humanity, then its task should be that of trying to understand the traditional and modern contexts which make these values meaningful. This means that religious studies is by its very nature not only cross-cultural but also interdisciplinary. In short, religious studies has links with every discipline inasmuch as it relates to the *humanum*, but while some links have almost eroded and ought to be reinstated, still others are far too overgrown and ought to be trimmed.

Out of such dialogue new problems emerge, new methods may be found and new hermeneutics invented. This process has already provided religious studies its own specialized group of scholars. Religious studies alone cannot adequately cope with all these fields and subtopics, which require abundant input from the other disciplines. The nuclear issue

is a particularly acute example here, but hardly the only one.
It is indeed methodologically wrongheaded to try to deal
with all these borderline issues in isolation, as if religious
studies had some sort of special dispensation or revelation
which would entitle it to make judgments about such
problems independent of their complex connections with
the disciplines concerned.

There is in this regard another holdover from former
theological methods that should stringently be avoided. It is
not adequate to obtain some results from the various sciences
and then mix them up in a religious studies cocktail. It is a
false assumption to suppose that we can really know the
results from a particular field without having participated to a
certain degree in the very methodology of that field. If we
cannot speak about a religious tradition without at least some
empathy and inner knowledge of that tradition, how can we
expect to do so for the other branches of human knowledge?
The methods are intrinsically connected with the results of
the study in question. Many astoundingly inaccurate ideas
about modern physics, for instance, would disappear
straightaway if those who think of elementary particles and
energy quanta as little material or energetical things would
get to know the methods by which modern physicists have
come to speak about these "entities." Religious studies is a
clearinghouse, I said, not a packing and baling depot. We
cannot deal with ready-made ideas. The dialogical character
of religious studies is central to its method.

It could be said that I am proposing to replace the old role
of philosophy, and that the religious studies faculty member
should be a kind of philosopher-in-residence in each
department. In a certain way it could also be said that I am
resurrecting the classical idea of the *studium generale*. Here I
should have to respond Yes, and No. Yes, since the university

has to be loyal to its name and not become a "multiversity" or a sheer "diversity." To avoid this, some agency or discipline has to try to bring together the different branches of learning and of reflection upon reality; it has to make the attempt, pluralistic though this may be, to assemble and integrate the dispersive vectors of human consciousness into the most comprehensive intellectual picture possible. No, on the other hand, since religious studies has neither a higher source of knowledge nor an a priori method, much less the power or the inclination to follow the lead of medieval theology and crown itself queen of the sciences or other disciplines.

* * *

The challenge put to religious studies today is to fathom and to face the full and sometimes terrifying implications of human freedom, draw the appropriate conclusions, act on them responsibly, and live with the consequences. Scott Eastham's *Nucleus* is an invitation to think and to act collegially —precisely in order to meet this challenge.

###

R. P.
Easter 1986
Santa Barbara, California

Sanity
is the most profound
moral option of our time.
Renata Adler

The Challenge: Life, or Death

Between the idea
and the reality
Between the motion
and the act
Falls the Shadow
T. S. Eliot[1]

The Pall of Oblivion

Today, as always, the sun rises and the Earth bathes in its radiance. Today, as always, human beings awaken to pursue their diverse activities, alone or accompanied, near or far from one another, some converging and others diverging, each a pilgrim on his or her own path. But today, as never before, an unseen shadow blankets the Earth, tarnishing even the bright promise of the morning light. Today, as never before, this impalpable shadow throws its pall over humankind, eclipsing every human horizon, stealing into every social interaction, darkening every pilgrim's path.

The shadow? It is the threat of terracide, of planetary

destruction and human extinction, an invisible shadow mocking the very light of day. It is the pall of oblivion cast by the existence of nuclear, chemical, and biological agencies of Death of a kind and on a scale unimaginable before our traumatized 20th century. And the weapons are only the most grotesque and appalling facet of global and ecological dilemmas without precedent in our individual or collective experience. We daily confront devastations of the Earth and her creatures as never told or foretold in the old books. Wherever we turn, the shadow of Death crosses our path: killer famines, political repression and exploitation, wars and persecution on almost every continent, pollution of the waters and the air and the soil, exhaustion of resources, extinction of species. . .the catalogue of contemporary horrors seems practically endless. Yet the upshot is that it all may come to an end, and very rapidly indeed. Omnicide or biocide—the killing of everything, the killing of Life itself, perhaps—may be even more appropriate ways to express the depth and magnitude of the abyss gaping before us today.

We live in an epoch of human emergency, a turn of phrase intended to cut both ways: we face on the one hand an extreme crisis—extinction of our own species and quite possibly Death of the Earth altogether—an *emergency* which admits of no easy answers. If we do manage to survive our present predicament, on the other hand, it will be because a new human reality has emerged, a new or renewed sense of what it means to be human on Earth and under Heaven. It is toward a realization of this *emergence*, this tentative new humanity, that the present book is dedicated. Nothing less than an anthropological mutation, a conversion of human being at its very core, may be adequate to the challenge of our troubled times.

What is the Nuclear Issue?

Albert Einstein's pointed remark of 1954 still stands as the most often cited summation of the nuclear predicament: "The unleashed power of the atom has changed everything save our modes of thinking, and thus we drift toward unparalleled catastrophe." The intervening 30 years only underscore the problem; atomic weapons have multiplied a thousandfold in numbers and in destructive power, but "our modes of thinking" have hardly changed at all. Where do we come by these modes of thinking, and why are they so stubbornly resistant to transformation? Presumably the universe at large is our most basic learning environment, the primary university. But most of us tend to pick up our ways of thinking critically about ourselves and the world from our early training in various educational systems. It is therefore legitimate to begin this discussion from our given situation as teachers and students in universities and colleges. Clearly, we are not the wide world, and we have parochial concerns of our own. But we have also a range of rather special resources from which we may draw as we bring ourselves face to face with the present human predicament in the nuclear age.

All our academic "modes of thinking" about teaching and learning today face the direct and uncompromising challenge of having to make sense of a world and a society which shows every sign of being hellbent on its own destruction. What is the meaning of education in such a world? Are we merely training people to function like mindless cogs in the mechanized rituals of the Death Cult which western civilization has by and large become,[2] or are we educating ourselves and empowering one another to discover life-enhancing alternatives to destruction?

If the question is focused this sharply, as the choice between ways of courting Death and ways of cultivating Life, then the nuclear issue inevitably becomes the *nucleus* of every academic enterprise, the core concern. Professor Shingo Shibata, formerly of Hiroshima University, insists that this question of omnicide or biocide radically reorders all educational priorities.[3] If history is about to end abruptly, for example, he would say that this looming dead end ought to be considered the central or "nuclear" issue for the historian. How has western civilization come to such a pass? Is there no alternative suggested by the millennial experience of human-kind? If society is about to annihilate itself, what other trends are more crucial for the sociologist to evaluate? For the psychologist as well, to study paranoia, schizophrenia, and psychoses of all sorts, while ignoring the larger sickness of an entire civilization that perversely persists in flirting with its

own destruction, would have to be the height of folly.
Further, if human existence is about to cease altogether, what
other existential issue calls more directly for the philosopher's
reflection? And if science is about to destroy not only the
humanities' budget in most colleges, but humanity itself,
then is the nuclear issue for scientists merely a matter of
reshuffling their computer data banks in order to procure
more government grants?

The questions are rhetorical; this is the paramount
human concern. Helen Caldicott once put it that, in the face
of total destruction, "Nothing else matters."[4] And as most of
us in religious studies are well aware by now, if religion
claims to concern itself with the ultimate salvation of
humankind, with the meaning and goals and values of
human life, then religious institutions can no longer turn
their backs on the concrete issue of human survival. Indeed,
religious colleges, universities and communities may not only
be able to provide a very special perspective on the whole
question, as we shall be stressing in the present book, but
may also find that they face a responsibility to do so. The
nuclear issue brings the problem of salvation directly down to
Earth. Whatever ideas of salvation the religions may wish to
promulgate, these will retain meaning only so long as human
beings survive to be saved.

The preeminence and centrality of the nuclear issue can
hardly be argued; it is literally a matter of Life, or Death, for
the entire planet. There is no more vital or more urgent item
on the human agenda. Even remaining within our academic
context, the problem can take on pragmatic contours. Are
universities merely to follow public opinion in this area, or
are we called upon to inform and inspire public opinion? We
find we can no longer ignore issues of such moment, or pretend
they are the professional bailiwick of some other specialization

than our own. But what then is to be done? How are we in
the somewhat rarified world of universities and colleges
going to come to grips with this issue of human extinction?

Well, one option would be to go on strike. We could simply
say to ourselves that the present arms race between the
U.S. and the U.S.S.R. is so awful and so out of proportion to any
human values that we cannot in good conscience conduct
business as usual any longer. We could, in short, stop
studying and stop teaching. Plainly, such a protest would
amount to no more than an impotent gesture of frustration.
It would also pathetically demonstrate our personal and
communal inability as scholars and intellectuals to deal
constructively with issues of such magnitude. It would ignore
the possibility that society as a whole may need and even call
for our guidance. And, ironically, such a strike could only be
contemplated if we had already written off our own academic
work as merely "academic" in the worst sense, that is,
irrelevant to the overriding concerns of a world in peril.

We could, on the other hand, do precisely what we know
how to do. We could begin to study this problem, and to this
study we could bring to bear all the resources—human,
textual, technical, administrative, and spiritual—to which life
and work in a university community provides access. Where
better to try to understand how humankind has backed itself
into this cul-de-sac, and to determine how such an impasse
might be surmounted? Happily, there has recently been a
groundswell of agitation and encouragement to engage
colleges and universities directly in this enormous project.
John Ernest poses the question in no uncertain terms: "What
are the universities doing and what can they do to arrest our
relentless drive toward cultural suicide?"[5] Similarly, the well-
known Peace Pastoral issued by the U.S. Catholic Conference
of Bishops in 1983 urged just such a full-scale commitment of

time and effort by all Catholics to the study of the nuclear issue, a move widely endorsed by other religious congregations.[6] So far, however, according to recent reports, there has been little follow-through on such suggestions.[7] Is our silence not part of the problem?

All such signs and portents underscore the critical need for institutions of religious education and programs in religion to address this issue. But how? How do we go about making the nuclear issue curricular, that is, a regular part of our programs of study? What in fact *is* the nuclear issue in religious studies? Do we know? More likely, we know what it is not. We know it is not only a technical or tactical problem; it is not for lack of proper tools or techniques that the super-powers have not dismantled their thermonuclear arsenals and, indeed, keep building more weapons. Nor is it only a military issue. All-out nuclear holocaust might well take little more than 45 minutes, rendering obsolete at a stroke all military contingency plans. With no victors, the military rationale for nuclear war is exactly zero. Similarly, this is not an exclusively political issue, or else we could expect a real program for disarmament to be evolved in political terms instead of constantly aggravated. After 40 years of negotiations (often a euphemism) in Geneva, it is all too clear that the politicians and diplomats are unable to resolve the problem on their own. In the context of the modern nation-states, national security will always tend to take precedence over human safety, rendering it highly unlikely that any purely political resolution will be found. Of course, technical, strategic and political concerns are salient dimensions of the problem, yet the dilemma resists being reduced to any one of its dimensions. The nuclear issue is not a single theme. It is a Gordian knot of human concerns, a complex of issues constitutively interrelated at every level of human intercourse.

But unlike Alexander, we today cannot peremptorily slice through the nuclear Gordian knot without also slitting our own throats.

Theology has an approach to the problem at the moral level, raising questions about just war and the pertinence of pacifism. But within a given theological framework, moral judgments are generally treated as decisions based on reason, and there's the rub. After 40 years of totally reasonable moral exhortations, and a concomitant escalation from three nuclear weapons in 1945 to more than 50,000 such weapons today, we might tend to agree with the poet Ezra Pound, who once ruefully remarked, "An appeal to reason is about a 13% appeal to reality."[8] But the field of religious studies is not bound by the categories of one theological schema or another. It looks to the assumptions and presuppositions of such metaphysical systems. From the perspective of cross-cultural religious studies, the formal (and usually Kantian) philosophical tradition to which thinkers on moral development subscribe is not grounded in an integral understanding of the human person, but all too often finds only a flimsy anthropological basis in rationality. Philosopher Max Scheler has written,

> It is no terminological accident that a formal ethics designates the person first as a "*rational* person . . ." With this one expression, formalism reveals its implicit material assumption that the person is basically nothing but a logical subject of rational acts, i.e. acts that follow these ideal laws (logic, ethics, etc.).[9]

In her recent assessment of morality and culture, Mary Douglas seems to echo this critique:

> Western liberal philosophy emerges historically as a series of refusals to accept any authority but that of

reason in its judgments of good and bad. . . For all
their disagreements, there is one thing philosophical
debaters are agreed upon: they do not even want to
know about the cultural formation of moral ideas.[10]

Most of us know from hard experience, however, that the
rationalist assessment of human character, stemming from
the so-called Enlightenment, falls woefully short of the mark.
Sad to say, even when convinced or convicted of logical
contradiction or a transgression of ethical norms, people do
not just automatically change their views or, more importantly,
their behavior. There must be more to it. Cross-cultural
scholar Raimundo Panikkar sheds a great deal of light on
such quandaries, which often appear opaque from a
monocultural perspective. We shall have recourse to
Panikkar's insights often in this book, for the extraordinary
leverage they provide on the entire range of dilemmas
comprising the nuclear issue. Here he would note that what
rationalism leaves out is precisely the shared *mythos* of a
given community and, by extension, all that "goes without
saying" in human affairs. Panikkar calls the myth that tacit
"horizon of intelligibility" which underlies the merely
rational criteria to which we appeal when a breach of right
relations becomes evident. As long as the common myth
holds sway, no moral decisions or judgments are in fact
necessary; this or that "just isn't done." But when the myth
breaks down, when the unthinkable becomes thinkable, or
when one human group encounters another group with a
totally different mythic orientation, all moralistic appeals to
reason or to law seem futile. Panikkar puts the situation as
follows:

> The myth of morals is morality itself, and when
> morals cease to be a myth, they also cease to be

moral. . . .We act morally as long as we do not ask why.
The moment we feel obliged to justify morals by
reason (and how else could we do it?), they begin to
crumble.[11]

One further point is in order, which cannot be rationalized
away. The nuclear issue is unavoidably an encounter with
Death, on both a personal and a species-specific scale. As
such, it is a primordially religious issue. It has often been
remarked that our attitudes toward Death condition our
attitudes toward Life, and so affect everything we do and say
and think and are. As Elizabeth Kübler-Ross aptly remarks,

> Death is the key to the door of life. . . . For only when
> we understand the real meaning of death to human
> existence will we have the courage to become what we
> are destined to be.[12]

The awareness of our own Death is our initiation into the
mystery of human Life; it is what makes us properly mortals,
we who know we shall die. In this respect, the megadeath
foreshadowed by nuclear weapons becomes our teacher; it
teaches us the inestimable, irreplaceable value of Life on
Earth. Initiation means a new beginning, but it is a beginning
that must be won, earned, conquered in the terrible encounter
with Death. From the most rudimentary shamanic initiation
to the resurrection of Jesus Christ, the religious life invariably
includes this experience of Death and Rebirth, in as many
symbolic forms and rites and mystical journeys as there are
religious traditions or, indeed, religious people. Religion is
concerned with the whole of human life, its meaning and
value, its beginning and end, and all the crucial stages in
between. Religions claim to offer human beings a way. . .to
fulfillment, salvation, justice, liberation, peace, and so on.

The anguishing question we must face today is whether or not the religions also offer human beings a way to survive this precarious nuclear age. How many religious organizations and religious leaders prefer to defend their narrow and sectarian views of religion even as their incessant "holy" wars threaten to catapult us toward planetary suicide? In short, what survival value has religion at the close of the second millennium C.E.? And if this most intractable question is not addressed, then all the good intentions in the world amount to no more than wishful thinking—or worse, escapism.

Human Relativity

For more than 40 years, humankind has been struggling to come to terms with the existence of nuclear weapons. As we have observed, the nuclear issue is at bottom an encounter with Death, with an extended horizon of our own mortality. We must unflinchingly recognize that the problem of human extinction is not going to go away. Even if we decide tomorrow to curb the production and deployment of genocidal weapons, our species is still going to have to live with—that is, survive—the ineradicable knowledge that we are capable of destroying ourselves utterly.

Yet if we do survive, it will mean that we have also learned from the nuclear predicament to value Life anew, to discover that Life, too, has a deeper horizon which never comes to light more clearly than in the encounter with Death. This more subtle dimension of human existence is our inter-dependence, our solidarity with one another and with Life altogether. Our very existence depends wholly upon a vast and intricate fabric of connections—cosmological, biological, personal, social, and cultural connections—which tie each human being intimately to the fate of every other. Either we learn to live together, or we succumb together.

All of this impels a radical change in focus. To be accurate we have to define, or redefine, the nuclear issue as the question of *human relativity* in all its dimensions. How are we human beings going to learn to relate to one another constructively instead of destructively, and quickly enough to forestall annihilation? Once you see it, the change in focus is an obvious one. Attention to relatedness—meaning right relations between people and peoples, not any kind of situational ethics or relativism—immediately unmasks the human face of the nuclear issue. It is first and foremost a human dilemma, one which reverberates through every stratum of human interaction, and simply cannot be confined to inventories of hardware or software or political opinion polls.

Seen from this vantage point, the nuclear issue naturally also refocuses our attention on many other crucial human relationships which are too often entirely bypassed in the political and strategic debates. Our relationship(s) with the Earth, for example, with the thousands of other living species that share this Earth, and all the interpenetrating ecological networks in which we participate—these too are part and parcel of the nuclear issue. Our relationship(s) with God and the Gods as well, our sense of our place in the scheme of things, our ultimate attitudes and priorities, the ideals we are willing to live or to die for—these primary relationships structure our entire approach to the nuclear dilemma. It is a custom at the end or beginning of Lakota (Sioux) prayers to invoke Grandfather Sky, Grandmother Earth, and *mitaoyate*— weakly translated as "my relatives," yet implying for the Native American not only our human relatives, or even the human family as a whole, but *all the relations* we share with every other living creature. Such a prayer rightly situates the human being between Heaven and Earth, at the crossroads of the entire reality, in the midst of these vital relationships which constitute our very existence. Only presented in this way do human difficulties appear with enough depth of field to permit their resolution or transformation.

Christian anthropologists would concur in this relational emphasis, reminding us that the nuclear issue has therefore a human nucleus, a personal center:

> Experience and most assuredly the Bible affirm that we humans are relational, as is all of reality. The basic ideas of the Bible—God, humans, love, hate, sin, conversion, fidelity—are relational at their core. I, as an individual, am not alone in this world; I am a person-in-relationships.[13]

Psychologist C. G. Jung, who sought the integration of the human soul in all its archetypal depths, decried the ways we ignore our personal responsibility for these constitutive relations and try to "educate" people to accept their own irrelevance:

> Scientific education is based in the main on statistical truths and abstract knowledge and therefore imparts an unrealistic, rational picture of the world, in which the individual, as a merely marginal phenomenon,

plays no role. The individual, however, as an irrational datum, is the true and authentic carrier of reality, the *concrete* man as opposed to the unreal ideal or normal man to whom the scientific statements refer.[14]

Jung understood more deeply than many theoreticians that in our day it may well be at this personal center that the destruction or renewal of the world will come to pass. When human existence itself is threatened, we ourselves may be our own last court of appeal. And in many respects, we human beings do not know ourselves (or one another) very well at all. We think of ourselves as isolated individuals and ignore the fabric of connections into which the threads of our lives are woven. We submit to the machinations of the powerful without recognizing that we have given them the only power they have, the power to delude us into thinking we are powerless. We shrug our shoulders at the evil ways of the world, without realizing that the world is the way it is only because we are the way we are. If we are to turn matters around, we have to turn ourselves around. Several closely related issues therefore call for our immediate attention—as scholars and educators, yes, but also and most profoundly as human beings.

The Response:
Religious Imperatives
of the Nuclear Age

What is most thought-provoking
in our thought-provoking time
is that we are still not thinking.
Martin Heidegger[15]

Prologue:
The Three Worlds
in Crisis

With obvious relish, R. Buckminster Fuller used to relate the following story about academic myopia. In New York City a few years ago, two learned societies met in separate hotels on the same day to discuss strikingly similar topics. The one group consisted of biologists attempting to determine the major factors causing extinction in species of flora and fauna. The other study group consisted of cultural anthropologists grappling with the causes of extinction among human tribes and groups. In both cases, the primary factor leading to extinction turned out to be the same: overspecialization. There is compelling irony in the fact that until it was pointed out to them a considerable while later, neither group of "specialists" had the least inkling of the other group's findings.

This little tale not only illustrates the nearly epidemic fragmentation of knowledge in our time, it also neatly underscores the possibly fatal consequences of compartmentalized scholarship. The moral of the story? If we're not careful, we may very well specialize ourselves right out of existence. It is not therefore amiss to adopt as our working assumption in what follows the principle that only an interdisciplinary approach stands any chance of illuminating the nuclear dilemma in all its interrelated aspects. But interdisciplinary study does not mean just comparing and contrasting data or texts from diverse disciplines, nor does it mean imposing on any one area of research methods derived from another. For our purposes, we may succinctly define interdisciplinary studies as the concerted effort to approach fundamental human questions from any and every available angle. The entire range of human experience is implied here—and not only horizontally, among human beings, but also as it were vertically, between what I shall be calling the three worlds.

It is almost a commonplace in the study of religion that human beings in most traditional religious societies inhabit (at least) three distinct and yet inseparable "worlds" or dimensions—mostly simply put, an above, a below, and an in-between. Ezra Pound paraphrases this threefold formulation from the Confucian classics: "Heaven, man, earth, our law as written/not outside their natural colour."[16] We might equally cite the *triloka* (the "three places") of the Hindu tradition, or Dante's *Divine Comedy*. In the shamanic worldview, which has in one form or another served as a basis for human culture at least since the Neolithic era, we consistently discover a depth dimension, an underworld of dark forces, inhabited by the demons of disease and by Death itself. Overcoming Death reveals a new life above, a height to achieve, a pinnacle, a world of light, the dwelling of the Gods or of God alone. And of course we always also find the human world, the daylight world of mortals, located somewhere in between and connected to Heaven and Earth by the great World Tree or *axis mundi*. These three worlds are the primordial dimensions of all human experience, but some people— notably shamans, mystics, poets, and seers—are a little more practiced at traveling between them.

R. Panikkar finds the three worlds to be very nearly a cross-cultural invariant in religious traditions worldwide, taking innumerable forms and expressions in various cultures. He stitches together the Greek roots for world (*kosmos*), God (*theos*) and Man (*anthropos*) and coins the word *cosmotheandric* to describe the reality constituted by these very basic cosmological, theological, and anthropological dimensions:

> The cosmotheandric principle could be stated by saying that the divine, the human and the earthly—however we may prefer to call them—are the three irreducible dimensions which constitute the real, i.e., any reality inasmuch as it is real. . . .What this intuition emphasizes is that the three dimensions of reality are neither three modes of a monolithic undifferentiated reality, nor are they three elements of a pluralistic system. There is rather one, though intrinsically threefold, relation which expresses the ultimate constitution of reality. Everything that exists, any real being, presents this triune constitution expressed in three dimensions. I am not *only* saying that every- thing is directly or indirectly related to everything else, the *pratītyasamutpāda* of the Buddhist tradition. I am also stressing that this relationship. . .flashes forth, ever new and vital, in every spark of the real.[17]

The following three chapters are intended mainly to register how the nuclear issue reverberates in each and all of these three worlds—the Biosphere, Humankind, and the Divine—as this entire cosmotheandric reality is confronted with the all-too-likely possibility of thermonuclear oblivion. We shall find first that no realm of the real is insured against catastrophe. But we shall also find that what is so deeply threatened is at the same time clearly revealed—the very fabric of connectedness crucial to Life at every level.

The Secular As Sacred

The past century has witnessed a profound shift in the
pertinent horizons of religious values. The long-standing
Christian segregation of the natural world from the supernatural has practically reversed itself; at the very least, the
formerly despised secular world has come into its own as
a locus of meaning and human commitment. It is difficult to
overemphasize the significance of this reversal, which may
indeed be the most significant turnabout in religious values
since the axial period (c. 500 B.C.E.) began to focus on the
human personality as an actor in history rather than on the
vast cosmological panoramas common to earlier, nonhistorical
traditions. The history of this gradual but all-pervasive

secularization is well known, as are the abuses of a reduction-
istic secularism or materialism which has all but lost touch
with the sacred.[18] But the secular is not the profane. Indeed,
the word *saeculum* means an age, a span of *time* (the Latin
version of the Greek *aion*), and therefore mainly stands for a
positive acceptance of time and history as evolutionary
processes.

It should quickly be added that secularization need not
undercut the magnificent primary thrust of the nature/super-
nature distinction—namely, that to be fully human is to be
called to a superhuman or divine vocation—but instead only
implies extending to the cosmos, to the natural world of
space, time, and matter, this selfsame divine abundance. In
other words, there is also *more* to nature, a bottomlessness
and spontaneity to every simple thing, a profundity to every
stone and flower, which cannot merely be reduced to the
mechanisms of empirical necessity. What is experienced here
is too often taken for a denial of transcendence, when it is in
fact a fresh discovery of the complementary spirituality of
immanence. Moreover, as the recent awakening of environmental and ecological sensitivities of all sorts testifies, this
spirituality of the immanent divinity harbors an internal
dynamism of its own. We seem indeed well on the way to
a complete sacralization of the secular, a sanctification of the
living cosmos, a new reverence for the world so long held at
bay in the two-storey dichotomy of nature and supernature.
It is, as Matthew Fox reminds us, a new point of departure for
religious awareness in the West: instead of concentrating on
the original sin by which we have fallen from grace, today we
seek ardently to rediscover and celebrate the "original blessing"
of the creation itself, in all its unceasing vitality.[19] R. Panikkar
provides a succinct overview of this new conversion to the
world:

The secular religiousness of our day. . . is in the midst of realizing the genuine experience of divine immanence. People devote themselves to the service of the earth, humankind, culture, society, science and even technology with the same *pathos*, the same seriousness, with which they formerly consecrated themselves to the service of God. . . . The spirituality of the immanent divinity makes modern Man fling himself into the arms of the World as into an absolute, as the immanent God he has discovered. Human salvation is seen as a liberation not of Man alone, but of the whole cosmos, as a liberation of the forces of nature, as freedom for the World as well.[20]

This shift in attitudes toward the world has also been felt in the natural sciences, and in the burgeoning attempts at a comprehensive and comprehensible philosophy of science. One need only mention Michael Polanyi, A. N. Whitehead, Jan C. Smuts, Thomas Kuhn, Lewis Mumford, Teilhard de Chardin, Buckminster Fuller, René Dubos, Lewis Thomas, Arthur Koestler, Lynn White, Ian Barbour, Thomas Berry, and more recently, Ken Wilber, David Bohm, John Haught, Brian Swimme, and others to demonstrate how far indeed we have come from the mechanistic reductionisms of the 19th century and into the organic world picture most of these 20th century thinkers have tried to paint for us. We shall be entering this new worldscape in a little more detail later. For the moment, it should be enough to note in the life sciences the emergence of new holistic models of symbiosis (instead of "nature red in tooth and claw"), synergetic structures and principles in the physical sciences (instead of the entropic running down of Newton's clockwork universe), and even a renewed interest in methods of homeopathic healing in medicine (instead of

outright surgical and technical intervention)—in short, a
renewed respect for the integrity of natural processes. Such
respect goes hand in hand with a sober recognition that all
living organisms exist within limits, regions, and boundaries
which cannot be violated without serious consequences. The
same attitude has also begun to transform the philosophy of
science: the subject/object or mind/matter dichotomy at the
bottom of so much technological abuse, has been of late
summarily overturned. One thinks of Heisenberg in physics,
Gödel in mathematics, Hofmann in chemistry, and Bateson in
biology. Even the once-sovereign principle of causality has
been sorely rocked by depth probes in subatomic physics
where time seems to jump backwards, as well as by the
preponderance of evidence supporting the existence of
various kinds of parapsychological phenomena.

In short, scientists are finding that they no longer can
regard themselves as detached spectators standing in splendid
isolation over against the objects of their research—detached
or, indeed, alienated from the entire natural world, which
had until recently all too often been conceived merely as raw
material for production and consumption. There is a new
sense of the ethical issues raised by even the most seemingly
abstract areas of research and a deepened quest for tools and
models appropriate and not inimical to convivial human life.
Midway through the 1980s, the edifice of science has also
become permeable, or at least more intelligible, to students
and scholars of religion and the humanities at large. The
ancient walls are crumbling; physics becomes metaphysics in
a flash, and also vice versa. As a culture, we are only on the
verge of exploring the implications of such changes, but we
would be well advised to do everything possible to facilitate
intellectual life on these frontiers.

The crucial interface here may be that between science

and religion, one of the least developed interdisciplinary areas due mainly to the vast number of incompatible specialized jargons that must somehow be reshaped in order for for the two to speak clearly to one another. Albert Einstein presented a powerful image of this often troubled relationship:

> Now, even though the realms of religion and science in themselves are clearly marked off from each other, nevertheless there exist between the two strong reciprocal relationships and dependences. . . .The situation may be expressed by an image: science without religion is lame, religion without science is blind.[21]

Teilhard de Chardin offered another image, equally compelling but perhaps a little more optimistic:

> Like the meridians as they approach the poles, science, philosophy and religion are bound to converge as they draw near to the whole. I say "converge" advisedly, but without merging, and without ceasing, to the very end, to assail the real from different angles and on different planes.[22]

Several clarifications are in order. To begin with, science is not a religion, and religion is not just a primitive attempt at science. They are quite different disciplines. Science prides itself on its rationality, but its roots run deeper, and in another direction. Science is the direct descendant of magic, which may be precisely defined as the prediction and (thus) control of empirical causality. Harvey Cox emphasizes the importance of this heritage:

The magical impulse is the desire to control and direct
nature, to use it for human ends, to tame its sometimes
malevolent side. This impulse developed through the
centuries not into religion but into empirical science.
The true successors of the sorcerors and the alchemists
are not the priests and theologians but the physicists
and the computer engineers.[23]

Scientific method is founded on the experiment, which is
verified only by its repeatability: anyone conducting the
experiment under the same controlled conditions will arrive at
the same results. Religion, on the other hand, deals with the
direct and unrepeatable experience of mysteries; its final
purpose is not to solve them but only to celebrate, that is to
say, participate in them. Science deals with problems; its final
purpose is to leave no stone unturned in its quest for answers.
Science is the modern gnosticism; salvation comes through
knowledge, predictability, certainty. But gnosticism is a
religious heresy. Religions by contrast generally hold that
salvation comes not by knowledge alone, but by faith in and
actions in accord with the spontaneity and freedom of the
spirit. In the final analysis, religion believes and trusts, but
does not entirely know what it trusts and believes (in).
Science, on the other hand, *knows*—(even *techné* originally
implied a way of knowing, "know-how") but believes in
nothing, except its own methods of testing: *subject versus object*.
    Today, at long last, we are beginning to face the fact that
technology has generated a crisis—the possible Death of the
Earth from biological, chemical, nuclear, or other causes—for
which there is no technical solution, no quick fix. We find we
can no longer take the Earth or her bounty for granted. The
revision of the theoretical relations between religion and
science as watertight academic compartments stems from the

pragmatic experience of discovering, in a thousand painful
ways, that the Earth has limits. R. Panikkar notes that unbridled
materialism—blind faith in an "expanding economy" for
instance—is really a theological idea gone berserk.[24] The
infinity of God is wrongly translated into an assumption that
the Earth, matter, time, space, and natural resources will go
on forever. But matter is not infinite, and we have had to
discover its intrinsic limitations the hard way, by famine,
scarcity, pollution, and exhaustion of resources. No more can
we assume that the air we breathe is suited for our lungs (not in
Los Angeles), or that the grass will continue to grow (not over
Love Canal), or the birds to sing (we lost the dusky sparrow
recently; entire species are going extinct all around us at an
accelerating rate). That these cosmic realities continue to exist
is nowadays dependent to a greater degree than ever before
upon human awareness of them, human attention to the
myriad ways these natural phenomena relate to one another
and to our human world. At this eleventh hour, we are
rediscovering our roots in the Earth, and our interdependence
with all creatures. Even the imperial "dominion" of the
Genesis creation account is lately being reinterpreted to mean
stewardship, and the "covenant" between God and humankind
revised so as not to exclude the entire creation. All of us are
more aware than our parents ever had to be that we live in a
fragile biospheric fabric of connection. The world is once
again beginning to look like a sacred place. . .

   Religion and science are clearly not a single discipline,
but neither can even be described in any depth without
reference to the other. The extremes meet, and embrace. In
theological terms, at issue here is the relationship between
God and the world. Are God and the world to be considered
one reality? No, or else we would expect this world to exhibit
everywhere a divine perfection, a divine goodness, a divine

harmony; and each of us knows all too well from daily
experience that perfection is not of this world. Some would
see in any attempt to perfect human life on Earth the hand of
God at work, but no one would identify God and the world
so completely as to suppose the project already finished. And
yet God and the world cannot be considered two, either. Two
what? Even to formulate such a proposition, we would have
to posit a principle or an idea or an entity superior to God.[25]
The ultimate and the proximate, the ideal and the real, the
metaphysical and the physical—in whatever terms you prefer,
God and the world cannot intelligibly be considered in isolation
from one another. What sense does the world make, apart
from its origin and destiny? And who is God supposed to be,
if not creator of the world? The one implies the other. An
integral vision is needed here, a nondualistic awareness
which neither subordinates the world to God or God to the
world, nor excommunicates these irreducible dimensions of
the real from one another. Such a vision cannot be prefabricated for mass consumption. It comes about as a uniquely
personal integration, the fruit of a spontaneous intuition. It
cannot be forced. It depends wholly upon the astonishing
capacity of the human psyche to assimilate and sustain its
experience of mystery, to tolerate the coincidence of opposites
beyond merely rational frameworks, to turn wrenching
dichotomies into creative polarities. This is the task of our
times, and it strikes to the very roots of our personal and
collective existence.

The threat of extinction is with us to stay; it is now an
ineradicable part of the human inheritance. It will exist as long
as we exist. Once the knowledge of how to make an H-bomb
exists, only an actual thermonuclear holocaust could expunge
the possibility of such a holocaust (that is, all the bombs,
books, and brains available to trigger it). And such knowledge

is readily available, even in the public domain of unclassified documents. Indeed, a high school course in physics could provide most of the requisite expertise to make a crude atomic bomb, if the materials were handy. Like the adolescent discovering that he or she can take his or her own life, most people learn to live with the possibility of suicide. But if one or two disturbed people take their own lives, the remainder of the human family is not threatened. In the nuclear age, however, if one or two people in the right place and at the right time fail to cope with the possibility of planetary human suicide, then we are all doomed. We have reached the unprecedented condition where philocide (or biocide) might very well recapitulate ontocide (or suicide).

It can hardly be overemphasized that the post-Hiroshima generation—my generation—is the very first to have grown to maturity with the terrible foreknowledge that any given human act could be our last act, and that the curtain could very well come down on the world theater of the human drama at any moment, never to rise again. An entire generation, well over two-thirds of the Earth's present population, has never lived a single moment without this sense of impending human and planetary catastrophe. It is not necessarily despair or fatalism that results, but a painfully acute sense of the dimensions of cosmic tragedy lurking behind the bland conventions of the current human comedy. If the religious experience par excellence is the encounter with Death and the emergence into new Life, or true Life, the breakthrough to another higher plane, then we humans have begun for the first time to confront on a species-specific level the selfsame religious archetypes which have governed personal and communal existence up to now.

One must moreover concede that the religions, however far and wide they may have been disseminated over the course of centuries, have in the main been fairly provincial cultural phenomena. Religions (even the "catholic" church) have until quite recently been more or less regional and local in emphasis; they are concerned primarily with the identity and the well-being of a single group or tribe or congregation or empire—people who choose to live in a particular way. Today, the religious dimension of the human being and of humankind itself is being challenged as never before to come to grips with a crisis that threatens the entire creation, the world itself. It is out of this soul-baring experience that a new sense of participation in all the rhythms and patterns of the cosmos is emerging. It is not emerging primarily at the level of doctrine, or ideology, or intellectual abstraction, but rather in an almost wordless awe of the divine "pneumatics," to use a medieval phrase, of the sacred character of the ebb and flow of all the dynamisms and energies that hold the world together. In the New Age, nothing is more sacred than the secular. There is a new or renewed feeling that the world makes sense, that it is a *cosmos* and not a *chaos*; an *ordo* or a *li*, an order or a pattern; a whole which coheres. We are finally beginning to discern in very precise terms a continuum, a gamut or spectrum of patterns which connects atoms to molecules to amino acids to proteins to cells to organs to organisms to people to communities to cultures. . .and, ultimately, to the communion of all living things in Life itself. This then is the pragmatic mysticism of the immanent divinity, of the holy spirit, of the *forma* that in-forms life at every level, of the presence of the whole in every nook and cranny of the real.

Science, helpful as it is in rendering such observations precise, can in the long run provide only one sort of perspective here. Etymologically, "science" is "scission," that is, what scissors do, cutting whatever it studies into smaller and smaller segments and analyzing the isolated fragments.

Science is primarily a method of knowing more and more about less and less, and this is its strength. All through the momentous early decades of this century, physicists sought the basic "building blocks" of matter—first the atom, which then turned out to have multiple subatomic components, these in turn held together by incredibly strong binding forces which we learned to unleash by bombarding nuclei with neutrons and other particles. Lately, ever more minute constituents of the nucleus—mesons, baryons, quarks and what have you—keep cropping up, but no "building blocks" have ever been found. We have in fact discovered no such "thing" as an atom or a nucleus, no ultimate monad or substance. All these supposedly elementary "particles" have turned out to be sheer relationship, as it were a radical relativity of energies interpenetrating at many levels or frequencies of articulation, and not things or solids at all.

And we ourselves are inextricably tied into the whole concatenation, as Heisenberg demonstrated nearly six decades ago: the very act of observation necessarily alters whatever is observed. So at the antipodes of scientific objectivity, a reversal comes about which throws us back into the all-too-human realm of subjective choices, judgments, and values. By scientific method we have learned how to break up atoms, but we cannot learn by the same methodology how to put a world of thermonuclear politics back together again. Even the exceptions prove the rule: in plasma physics (fusion) and the recent breakthroughs in recombinative DNA (gene-splicing), scientists are beginning to put things together instead of destroying whatever they study, but we know so little that the problem is not solved but redoubled by the introduction of fusion bombs and (soon enough) mutant genetic structures into our terrestrial biosphere. Plainly, a more integral approach is required.

The splitting of the atom, the *atomos*, once considered "indivisible" as its etymology indicates, represents a new dimension of human damage to the universe: terracide literally means "the killing of matter." Once the integrity of the atomic nucleus is violated, once the atom is forcibly smashed, we now know that this produces something like a rip or a tear in the very fabric of matter. Errant neutrons are scattered about which collide at random with other nuclei, breaking these apart and setting off a chain reaction which never really stops. The affected nuclei often go on decaying for millennia. And when the radiation they emit strikes living cells, these nuclei too disintegrate or run amok, producing the organic chain reaction we call cancer, an unchecked local explosion of cell growth which eventually kills the host organism. Cancer is doing to human beings exactly what human rapacity has long been doing to the entire biosphere. There is irony here, perhaps even retributive justice, but scant consolation. The process of learning that scientific detachment is a pernicious illusion has also taught us that there is no pure science—once upon a time presumed free of moral judgment, and therefore in effect often quite impure. Here a real distinction ought to be maintained between research (science proper, the open-ended inquiry) and development (technology as systematic application), a topic to which we shall return. In short, we find we must take responsibility; we find that our very lives are implicated in the fate of the world, and also conversely.

The physical structure of reality has revealed itself to be event-ual, act-ual, and dynamic, a living tissue of inter-dependence and not a static set of building blocks, either physical or conceptual. Einstein postulated the convertibility of matter into energy ($E=mc^2$), enabling us to break into the heart of matter and release the very forces which hold the

atomic nucleus together. We have forgotten that the equation
also works in the other direction: that energy materializes,
that light congeals, that awareness takes shape, that the word
becomes flesh. In short, we have forgotten how to go about
putting the world back together again, and at any moment
now it threatens to burst or collapse. At this extreme point of
no return, it would seem, the healing vision of the whole
begins to surface in many disparate disciplines. To heal is to
make whole, and the common Anglo-Saxon root *hal* reminds
us that this hale and hearty healing that makes whole is
precisely the work of the holy.[26]

It is not therefore from caprice but from survival necessity
that the need arises with ever more urgency to reassess the
relations and rapports between religion and science, to build
a sturdy common language between these two universes of
discourse—between God and the world. What holds the
world together? How can we help maintain the existence of
the very cosmos in which quite literally and physically we live
and breathe and have our being? Can we have a creative or
re-creative relationship, instead of a destructive and ultimately
self-destructive one with the creatures and indeed the entire
creation of the creator? In so many words, we need a new
cosmology, a new or renewed vision of the whole—not from
a single perspective, but from every vantage afforded by the
diverse cultural, religious and, surely, scientific traditions.
Fortunately, a great deal of very exciting work has lately gone
into spelling out this new story of the cosmic, human, and
divine adventure which Thomas Berry, John Haught, and
others have so eloquently heralded.[27] Yet a new cosmology
will not be the vision of one man or one woman. Nobody has
an adequately universal perspective; it will have to be a
collaborative effort. A new cosmology might not mean merely
a new description imposed upon the world, it might mean

this very world inhabited in a new way. It will take whatever
shape it takes directly in our midst, a world continually in the
throes of birth: *cosmogenesis*. And it will re-form us as well: if
we humans are too often the abyss between God and the world,
we are also the only conceivable bridge. Perhaps Einstein's
Unified Physical Field Theory was stillborn because it was
only physical, leaving out the crucial human and spiritual
ingredients. Post-Einsteinian metaphysicians would be equally
bereft were they to settle for being only "meta-". There are no
disembodied spirits, and there is no way to divorce metaphysical principles from at least some cosmological
assumptions. Reality only exists to the extent it is realized,
embodied, incarnated. And in the nuclear age, our vision of
the physical cosmos must encompass not only its opposite, as
Jung emphasized with regard to psychic integration, but also
the possibility of its own outright annihilation. These are the
signs of our times.

The Human as Humane

Who has the voice, let alone the right or ability, to speak for the entire human family? Who among us, knowing his or her own shortcomings, would dare to speak one for another? Nobody can claim a monopoly on human nature or experience, let alone on truth, beauty, goodness, or God. Humans are not merely the objects of any cultural, philosophical, or religious inquiry. Humans are also, and primarily, the inquiring subjects—subjects who love, suffer, hate, and hope as well. To know who we are, we have to ask one another. To know one another, we often have to ask ourselves questions we would rather not raise. Human beings are these peculiar creatures whose understanding of themselves and one another is part and parcel of their very being. Alone, no one, no person, no group, no nation, no single ideology or culture or religious tradition has the whole answer to the nuclear dilemma. But together, living and working and loving together, listening to one another, criticizing one another, even irritating one another, we human beings hold in our hands whatever answer to the current human predicament will ever be forth-coming. There is no other answer than a (possible) human consensus, difficult though this may be to envisage given the desperate and deplorable state of international "relations" in the current global military and political theatre.

The emerging cosmology evoked earlier is beginning to provide us a way to find ourselves at home in a meaningful universe, as well as a concomitant sense of responsibility to and for this living planet. Just as urgently, we need an integral anthropology, a new way for human beings to understand themselves and one another. Echoing Panikkar here, I would go so far as to call for a cross-cultural and inter-religious anthro-pology of mutual understanding,[28] which is a very tall order indeed. The important point is that this must needs be a *dialogical* anthropology; the human story is not a monologue to be told from one supposedly omniscient narrator's point of view, but a mutual undertaking, an ongoing and often entirely unpredictable conversation.

It ought to be evident that as long as we continue to relate to one another only as adversaries, as long as we have recourse only to conflict models of human relativity (politics, legalism, dialectics, games-theory, war), so long will we continue to squander whatever time we have left on this Earth fighting one another and our common habitat. If we continue to insist on viewing all human relations in terms of us versus them, the world we construct on this basis will inevitably tend to tear itself apart. We have nuclear weapons because they do.

We are preparing to destroy them because they are preparing to destroy us. Who are they? Who are we? Same difference. It is by now a psychological truism that human beings tend to project their own worst tendencies, their own propensities for evil, onto some convenient other. But as C.G. Jung pointed out as early as the 1950s: "It is the face of his own evil shadow that grins at Western man from the other side of the Iron Curtain."[29] The gauge of how little we know about ourselves is how little we know—or care to know—about the other human beings with whom we must learn to live (or die trying). In simplistic terms, war is but a magnification of this ugly human incapacity to cope with the presence of other human beings—an incapacity nowadays writ large, and threatening the very existence of human life altogether.

It is already plain, then, that an integral anthropology cannot be set forth in vacuo as a finished system, or a doctrine, or even a specific series of methodological procedures. Dialogue does not occur in the abstract. It begins and ends in the concrete daily encounters we human beings have with one another, and in the creative spontaneity each of us brings to these encounters. It can however be sketched as a programme, a collaborative project or, as it is styled here, an agenda for collegiality. What are the anthropological premises for such an agenda?

There is no abstract "humanity": only people, each unique, each unrepeatable and irreplaceable. "Humanity" has no firm referent as a noun if it is supposed to denote a demographic entity; its borders would have to shift every moment as people are born and pass away, leaving the generic "humanity" a vague and ambiguous non sequitur. Each cell in the human body, moreover, has a definite lifespan; thousands of cells die and regenerate daily. None of us even consists of the same physical stuff as we did a year ago. We are all in process, undeniably; but just what kind of process is this? Humanity can and should be more clearly defined as "kindness," that is, as an act, the nominal form of a verb, indeed, a way of life: precisely (the) *being humane*. Now granting the obvious— namely that people are not always good or kind or loving— also requires us to recognize that we have not yet attained our full humanity. Hence we shall occasionally employ the neologism "human(e)" to indicate the paradoxical situation of the-human-being-not-yet-fully-humane. This shorthand is mainly intended to emphasize that being human(e) is not a *fait accompli* but a pilgrimage; not a birthright but a constant struggle; the human being that being, as Heidegger puts it, whose being is at issue for it: a question, a quest for what this same thinker goes on to outline as the "care-structure" of human(e) being.[30] In short, *"homo viator,"* as Marcel phrased it: being human is being on the way.[31] I would go so far as to say that humanity really only exists to the extent that we are humane, and thus would opt for human*kind* as the generic noun which best brings out this kindred character of human(e) being. Such a definition, in its recognition that we are not yet either wholly humane or wholly ourselves, tends to underscore the religious dimension of human life—the un-finished or in-finite dimension.[32] Along these lines Mircea Eliade, in his recent autobiographical conversations, bids us to "conside that every human existence consists of a series of initiatory ordeals or trials; man creates himself by means of a series of conscious or unconscious initiations."[33]

All of the above represent only a quick sampling of motifs familiar to most of us in religious studies by a hundred different names—being and its becoming, initiation, transformation, regeneration, and so on. The contexts differ, but in all of them we find the human person at center stage as the crucible of its own transformation: being in the making. It goes almost without saying that personhood so understood takes root not in the isolation of individuality and egocentricity but in the fullness of human(e) relations. As we have already observed, these constitutive relativities of the human person trace through and bind together the entire reality of our experience; Heaven, Earth, and all human intercourse. Although traditionally giving short shrift to the Earth and all the "lesser" links in the Great Chain of Being,[34] Christianity at its best has always hallowed the primacy of the human person as a reality not to be annulled or supplanted by any other cause, principle, sovereignty, or ideology. And the Christian understanding of the person is as a reality constituted utterly—that is to say created, nurtured, and redeemed—by love. Such an understanding might very well serve as one of the basic tenets of an integral anthropology.

The nucleus of the nuclear issue, as pinpointed earlier, is the way we human beings choose to relate to one another, to the whole of creation, and to its source. It is therefore nothing short of a refusal of our human(e) vocation that leads us to put all our relations on adversary bases, to build conflict models, 180-degree antagonisms, dialectical strategies and win/lose, us/them, black/white, right/left, profit/loss oppositions of all sorts into everything we do together as human beings: wars literal and intellectual, pecking orders in every conceivable kind of bureaucratic and managerial hierarchy, competition in business, legal dialectics, sporting activities, educational programs—everywhere you may care to look. Just as we have squared off against the Earth as if she were an enemy or a competitor, so too have we put ourselves at odds with one another as long as anybody can remember: *subject versus subject.*

We shall reserve for later a more detailed discussion of organized warfare, which seems to be a perversion endemic to the rise of divine kingship, sovereignty, and civilization. For now, it is enough to note that neolithic village communities did not and to this day do not as a rule engage in the sustained mass slaughter we know as warfare. Such villages are locally ruled, fairly autonomous, and un-citi-fied, which early on meant mainly un-forti-fied. On the other hand, our own "civilized" history, that of city-culture, is notable mainly for its unmitigated savagery. We choose to forget that an entire society has to be bent on Death for the mechanisms of Death to be as successful and efficient as those we moderns have devised to do damage to one another. In the 20th century, we have lost our illusions about war: all our once grand fantasies that war will bring us paradise, utopia, peace, democracy, freedom, the classless society or the affluent society, today ring hollow. The nuclear issue has begun to demythicize war. We now seek only to "get by", to survive. But to survive what? The suicidal tendencies of our own species? Or perhaps these tendencies are really specific to modern western culture, and are only now threatening to spill over and engulf all the suddenly westernized peoples of the Earth. Is there no alternative to the fire next time? Are we going to let *homo maniacus* scribble the epitaph for *homo sapiens*?

While it is certainly true that every great religious tradition has outlined viable alternatives to conflict, it is also undeniable that most of the major religions have just as readily provided occasions for conflict, justifications for collective bloodshed, rationalizations for slaughter. Buddhism may be the only major exception. Too often and too easily has the blessing (*blētsian*[35]) been made with human blood. It is almost a platitude to observe that religion brings out the best and the worst in people, but are we still unable to separate the wheat from the chaff? For every Gandhi, there seem to be a score of Jim Joneses and Ayatollahs and Jerry Falwells sounding the first or the last trumpet in their never-ending "holy" wars.[36]

We shall examine later the extent to which warfare itself, as collective human sacrifice, represents a deeply-rooted aberration of religious values and practices. For now, it is enough to stress that both the origins of organized warfare in the cradle of the historical civilizations and the putative obsolescence of war in today's nuclear arena must be considered religious questions of the first order.

In the context of Christian moral teaching, these questions have customarily (and most recently in the American Catholic Conference of Bishops' 1983 Pastoral Letter, "The Challenge of Peace") been posed in terms of two distinct traditions, *pacifism* and *just war*, which taken together seem to reflect the ambivalence of a Church which has included both martyrs and soldiers, and which over the course of two millennia has preached both love for the enemy and crusades against him. Both moral traditions are sufficiently well-known that I may forego elaborating details in the development of these doctrines that are readily available elsewhere.[37] But the present nuclear predicament is different in kind from any other situations these doctrines have ever had to face, and a few remarks may therefore be pertinent at this juncture.

Just War?

The time for equivocation is long past. *Nuclear holocaust is not just war.* It is neither war in any accepted sense (no winners, all losers), nor can it be justified as a means to any human end(s) whatsoever, unless this be the end of humankind altogether. The three most prominent pillars on which the traditional Catholic "just war" doctrine has rested since Augustine—self-defense, proportionality of means to ends, and immunity of noncombatants[38]—have all buckled in the face of the nuclear arms buildup. The only self-defense against annihilation is prevention, Star Wars fantasies notwithstanding. Further, as noted, there are no ends which can justify modern nuclear, chemical, or biological weapons as a means. When the means will render the planet uninhabitable, the very concept of ends is rendered meaningless. And the so-called immunity of noncombatants seems little short of a bad joke given the destructive power not only of nuclear weapons but even of what we too facilely call conventional weapons. We tend to forget that until World War II, war had been somewhat tamed and its carnage generally limited. That is to say you had uniformed teams of young men, roughly equal in age and force of arms, battling it out on circumscribed battlefields carefully segregrated from populated areas. Certainly errant bands of soldiers laid waste, as they used to say, to towns and villages, burned the crops, and raped the women. But the intent was almost never simply a blanket extermination of civilian populations. The Nazi blitzkrieg and the Japanese massacre of Chinese civilians profoundly changed our conceptions and conventions about how wars were to be fought. Before the war we loudly denounced such atrocities, but by war's end the Allies had adopted (for example, in the scrupulously engineered firebombings of Dresden, Leipzig, and Tokyo) the same extermination policies we had decried in the Nazis. When the atomic bomb came along in 1945, it could therefore be viewed as simply a more efficient means to achieve this mass slaughter of civilian populations (or "countervalue strikes," in today's specious jargon) and was widely and easily accepted without major qualms of conscience as merely "a bigger bang for the buck." But the just war doctrines of moral theology are not so easily modernized; they stem from a deeper and, let us hope, more durable tradition. In the straightforward terms of this moral tradition, it is arguable that modern military scenarios— nuclear or conventional—can no longer even be called "war."

It is instead genocide with which we have to reckon,
wholesale people-killing, indiscriminate slaughter of
innocents, and it is anathema. Although hedging on the
question of nuclear deterrence (for which there is little
precedent in the tradition), the Bishops' Pastoral clearly
recognizes that no use of nuclear weapons can ever be
morally justified. Since Hiroshima, plainly, any attempt to
justify the use of nuclear weapons on human beings—and,
indeed, even the signaled intention to use them, which is
part and parcel of deterrence policy—necessarily implies the
demise of any ethical or moral system which would
countenance such a bald rationalization. It is perhaps less
obvious that the superpowers do use nuclear weapons as
instruments of policy every day, and have been using them in
very threatening ways since 1945, for exploding them is not
the only way to use such terrible weapons. They have been
used repeatedly by one human group to blackmail and coerce
others, much as a revolver may be used to rob a bank, even
though the trigger is never pulled. Nuclear weapons inspire
terror by their very existence, and in this sense the U.S. and
the U.S.S.R. have become the foremost terrorists in the world
today: between them they hold hostage the Earth's entire
population. The Peace Pastoral was a high water mark
because it reminded us that our moral traditions, though
sorely shaken by the current nuclear predicament, and
sometimes slow to react, can still help to unmask all the
pretense and overblown rhetoric and are not utterly bankrupt
in the face of an unprecedented situation. Whether they will
be heeded is another question. Before and after the bishops'
laudable effort, there have been plenty of counter-efforts by
various interest groups to turn the just war principle on its
head, using it to justify the deployment of nuclear weapons
which by their very nature violate its most basic tenets.

Pacifism?

Christian nonviolence has a venerable history in the life
and teachings of Jesus Christ and in the pacifist lifestyles
espoused by some early Christian communities.[39] But as the
bishops note in their Pastoral, and despite endorsement by
the Second Vatican Council of conscientious objection as a
valid Christian position, "The nonviolent posture in the
history of the Church has always posed a substantial challenge
to the whole community."[40] To say the least. In the nuclear
age, this challenge has come home to Christians as never
before, but to simplify our discussion we should emphasize
the distinction between nuclear pacifism and pacifism of any
other sort. There is a difference in kind and not just in degree
between the use of nuclear weapons and the use of force
when, say, confronted by a mugger in a dark alley. Still, we
should not minimize the consequences of the pacifist option
in the nuclear arena. Let us say I am a committed pacifist. Let
us say I am convinced I have an venerable moral tradition
supporting my stance, or perhaps the more recent moral
imperatives enunciated by Gandhi, or Thomas Merton, or
Martin Luther King. For me, unilateral disarmament is not
only the sole credible Christian moral posture toward nuclear
weapons, but also the only feasible first step toward
omnilateral disarmament. I should point out here that
disarmament must need be more of an attitude, a process, and
a gesture, than a feat which can be accomplished overnight.
Even if we managed to dismantle half a dozen nuclear bombs
every day for the next ten years, we would at present count
still have two or three thousand such weapons left over—
enough to incinerate every major Russian city and military
installation. But let us say that I insist we make a unilateral
start at this enormous project of disarmament. I believe that
mutual trust must start somewhere, and I am personally willing

to take the risk. As I have outlined it, this position is not
negotiable for me; I cannot in good conscience compromise one
whit, and I am willing to lay down my life for my conviction.
Let us even go so far as to say I am right. But what happens if
you, following the dictates of your own inviolable conscience,
sincerely disagree with me? What if you believe, to the
contrary, that it is the deterrence policy of mutually assured
destruction that has kept us out of a nuclear conflagration for the
past four decades? Even if I can in good conscience sustain
the enormous assumption that I am right and you are wrong,
by what moral authority could I then turn around and claim
to impose my opinion on you? If I really believe in a totally
nonviolent posture, how the devil can I justify forcing my
position on others, and requiring that they too stake their
lives on it? The dilemma is agonizing, at both the personal
and the collective levels, as the often heated debates between
some of the bishops clearly demonstrated. But here is
precisely where the transformative power of nonviolent
witness, noncooperation with evil, and civil disobedience
comes to light. One seeks to persuade, not to conquer. One
acts in a nonviolent way without regard to the calculable
effects of such action. Nonviolent witness has meaning only
within the context of faith: the motivations for your actions
stem from a deeper source than the arena where your actions
are played out. Otherwise, if good and evil are construed to
be on the same level, merely equals, slugging it out over the
eons like Ahriman and Ohrmazd, then unceasing conflict
will be the only outcome. Prayer, it is said, is the art of the
impossible. Only if I am somehow able to tolerate the
intolerable, that is, the existence of people whose views do
not coincide with my own, whose actions or doctrines or
ideology may indeed be totally incompatible with my
own . . .only then does nonviolence make sense and bear

witness to another order of things. It is thus a mystical tolerance
that is called for here, in the authentic sense: a tolerance that
sublimates what it tolerates, as the Christian tradition of patience
might still teach us.[41] The practical problem of nonviolence in the
nuclear age is at bottom the question of planetary pluralism:
how do we deal with truly incompatible worldviews, ideologies,
religions, cultures, and economic systems of all sorts? In this day
and age, we know all too well that we cannot annihilate our
opponents without at the same time courting our own
destruction. How, then, are we to arrive at a positive understanding and acceptance of human diversity?—one that we can
live with, so to speak.

* * *

Permit an abrupt assertion: the human reality is neither
monistic nor dualistic. Humankind is not one of anything,
nor can anything human be reduced to absolute unity. The
human world is not monolithic, monolingual, monocultural, or
even, on balance, monotheistic. Any attempt to reduce the
human reality to one—be it one empire, one religion, one truth
or whatever—invariably has to come to terms with the other.
As soon as you have fabricated your one world, the other can be
discerned hovering just over the horizon. We are not them.
Our group is human, and all the rest somewhat less so. At first
the others may seem easy to dismiss. They are just strangers,
foreigners, aliens, and we know next to nothing about them. But
as soon as they assert themselves in some way, the others begin
to look dangerous. They are soon enough the infidels, the
barbarians, the heretics who would undermine the unity (today
read: national security) of our society. The other looms larger and
larger, and in no time metamorphoses into the enemy. Just as
monotheism always generates its devil or its Antichrist, so the

one-world-empire has always to have its enemies of the state, be they external or internal, real or hallucinatory. In the final analysis, the other seems so threatening that the one may well destroy itself in the effort to ward off its implacable nemesis.

Contrariwise, much as we are encouraged to suppose so, humankind is not two of anything either: two parties, two factions, two ideologies, or two world powers vying for hegemony. It is no doubt easier to divide the world into good guys and bad guys than it is to face the true diversity of the human reality, or indeed, to seek a middle ground. But just as monism turns out to be vulnerable to its own shadow, so any dualistic attempt to divide the human family into two camps or two parties soon succumbs to the pitiless logic of the arena, where all the combatants only find themselves reduced to the lowest common denominator. If outright conflict is avoided, then very quickly a sort of terminal blandness tends to set in, a stale uniformity and routinized predictability—so much so that, in our dualistic political system, for instance, the one candidate for president begins to look and talk suspiciously like the other candidate or, in big business, the one corporation's products and markets come so closely to resemble those of its competitor that they finally merge, and so on. Six of one, half dozen of the other: a stupefying monotony insidiously infects any prolonged dialectical division of the human world. The two become locked into their struggle for supremacy and end up mirror images of one another, two heads of the same monster.

There is an analogy between human relativity and the principle of complementarity in nuclear physics. Just as the atom turned out to be neither wave (one) nor particle (many) exclusively, so the human world is neither one seamless fabric nor many separate worlds. It is all relative, and constitutively so. Thus no single perspective or description or explanation of human being(s) can ever be adequate. Unless or until the last human voice has been heard, nobody has a monopoly on human nature or culture. Hence the suggestion above, that a true planetary pluralism (which is not just an unrelated plurality) can come about only through a positive acceptance of human, cultural and—need we add?—religious diversity.[42] In human terms, just this is the nuclear issue: how do we deal with the other?—not merely as an object of our plans, calculations and strategies, but as a knowing and feeling subject, as our own necessary partner in the co-creation of any truly human(e) civilization on this Earth.

Behind the arcana of quantum mechanics, behind the technical ins and outs of hardware and software, behind all the political maneuvering and posturing lies this problem of human relativity, still largely unrealized. Neither monism (a hot war; annihilation) nor dualism (a cold war; stalemate) will provide pathways out of this morass. A true pluralism, simply stated, seeks ways for human beings to relate constructively to one another. To borrow a term from Gregory Bateson's reflections on biology, what we sorely need is a balance: a healthy cultural "co-evolution,"[43] where changes in the self-understanding of one human group set the stage for changes in an other, which in turn encourage further changes in the first group, and so on, reciprocally ramifying and reinforcing any newly emergent human(e) reality. Donald Keys points out that the Russians have coined their own term for this kind of reciprocating relationship; they call it "the policy of mutual example."[44] Unfortunately, some sort of co-evolution occurs whether the example set is a good one or a bad one. In the nuclear arena, bad examples predominate: the historical trend is toward incremental escalation of weapons. Charles Osgood has sensibly proposed turning the process around in what he calls a "calculated de-escalation":

> It is a strategy in which nation "A" devises patterns of small steps, well within its own limits of security,

designed to reduce tensions and induce reciprocating steps from nation "B." If such unilateral initiatives are persistently applied, and reciprocation is obtained, then the margin for risk-taking is widened and somewhat larger steps can be taken.

Both sides, in effect begin edging down the tension ladder, and both are moving—within what they perceive as reasonable limits of national security—toward a political rather than a military resolution. Needless to say, successful application of such a strategy assumes that both parties to a conflict have strong motives to get out of it, which obviously should be the case for both superpowers in this nuclear age.[45]

In theory, such a constructive co-evolution ought to gather momentum at least as readily as the massive cultural de-volution we call the arms race. In practice, there is often fear and resistance to even the smallest steps in this direction. The Russians under Secretary Gorbachev have indeed recently been making commendable and apparently quite serious overtures toward a comprehensive test ban and a "calculated de-escalation" of weapons levels, but such overtures tend to be routinely turned down by the U.S. Defense Department. Why? As a general rule, until cultures understand one another sufficiently to ease their mutual paranoia, common sense will always capitulate to exaggerated suspicions. Other animal species have learned to co-evolve; why should human societies, even faced with their own extinction, find the transition to co-operation so singularly difficult? What is it we do not yet understand about ourselves, about one another?

* * *

In point of fact there is a wide-open universe of discourse where we may try the mettle of our tolerance and test our

understanding of one another without hurting anybody. Every
society has spoken its own words, painted its own pictures,
and understood itself in its own way. Hermeneutics has always
stood for the art of interpreting the text(s) of a tradition to
apply to present-day exigencies. The term has classically
been reserved for legal and theological texts and their retrieval.
In academia, lamentably, this art of interpretation has too
often degenerated into theories of interpretation with negligible
issue; but it should be obvious that an exclusive preoccupation
with method will produce no results. In the cross-cultural
context of today's global crises, however, hermeneutics can no
longer mean merely updating one's own tradition, or elaborating
yet another theory of interpretation, however noteworthy
such efforts may be in their own precincts. Today we also need
what might best be called a *cultural hermeneutics,* an attitude
of mind and of heart which seeks to understand and
constructively relate disparate loci of ultimate human values.
Much of what passes for national policy in the nuclear age is
a collective failure of the imagination, a failure rooted in our
seeming inability to understand other cultures according to
their own lights. It is in this sense that we sorely need to meet
on their own terms all the great fictions, all the guiding
visions and grand fantasies by which the people and peoples
of this Earth have oriented their lives. This is not the place for
an excursus on the "myth of history,"[46] though it should by
now be patent that we have been reading the human saga from
one very limited (historical, future-oriented, objectivist,
imperial, linear, western, etc.) point of view. If our human
heritage is supposed to be only "his-story," as the feminists
say, then it is obvious that we have yet to hear the whole
story. And this is not a luxury, but a necessity for constructive
dialogue. Even comparative literature is insufficient to
encompass what is so urgently required: a thoroughgoing

rereading—and retelling—of the human story in its entirety.

During the past century or so nearly all of the world's classical religious texts have been translated into modern languages, and into English perhaps most of all. But these texts exist in our language mainly as literary artifacts, editings, selections, and translations which are usually the work of a single author and all too often bereft of the cultural and cultic context which originally gave them life and meaning. World literature is perhaps our most direct contact with the lived experience of other people, other cultures, and other religions. We neglect at our peril this living palimpsest of human attempts to render the mystery of human existence intelligible. We too easily tend in our literary criticism to stay at home, content with the convenient genres and conventions of our comfortably shared social fictions. Only if we do not flinch from encountering the truth insofar as it reveals itself in literatures founded on bases other than those of mainstream western civilization are we ever likely to discover viable alternatives to the increasingly apparent destructive tendencies of our own civilization. Nearly all the classics of world literature are religious texts, many of them "authorless," but we have forgotten how to read them as anything more than quaint exercises in literary style, period and genre. Following George Steiner, the basic hermeneutical dimension of religion and literature as I am sketching it here would stem from the principle that all understanding—within a given language, as well as between languages—amounts to translation.[47] The texture of human experience as a whole is to be found in the interplay between text and context. It is all a matter of interpretation . . .

If the goal we seek is dialogue insted of holocaust among conflicting viewpoints in the world theatre, then the only appropriate means of study will themselves display a dialogical character. This is what I would like to call *the dialogical imperative*. What does it mean? The goal and the means to the goal must not be incompatible. In the classroom, for instance, I hope later to show that there are still so many open questions that the role of the teacher must necessarily be less that of an authority who claims to have all the answers and more that of a facilitator who helps the students to inaugurate a genuine dialogue on the issue. We are all too practiced at taking sides and proving (to our own satisfaction) the rightness of our views, at choosing up teams or factions and battling it out until somebody inevitably loses. But we are awkward and inexperienced novices at constructive dialogue, where each partner assimilates and builds upon at least some of the insights of the other. And this inadequacy tends to afflict our pedagogical methods as much as it continues to stymie our attempts at arms negotiations.

* * *

To summarize: a religious anthropology, that is, a cross-cultural anthropology of mutual understanding, will remain a pious platitude until we learn to take seriously the self-understanding of other people, cultures and religious traditions. Reading and writing are presumably the interdisciplinary disciplines par excellence. Literature is not only our contact with the humanities in the academic world, but with humankind at large and the roots of human memory. One might also find here a Christian imperative: I am to love the other as my self—not as the object of my study, but as a "me" who is different than I am, as a unique and indispensable source of wisdom and under-standing. Anything less than the sincere attempt to understand the others as they understand themselves is worse than mere reductionism, it is really the vestige of a colonialistic attitude which would uncritically and perhaps even unconsciously subjugate the other under our own concepts of whom he or she or they may be.

The Divine As Mystery

The American poet William Carlos Williams once put it that "everyone should know EVERYTHING there is to know about life at once and always. There should never be permitted, confusion—".[48] The Efiks of Nigeria say that one "has no ears" until they are pierced in the rites of initiation, where one is told the whole story of Life, Death, and every step in between, in words that may not be written. The Christian tradition will remind us of the unceasing creation and redemption of the world at any given moment, and in every mass: the presence truly alive in every authentic present. The Vedic lore affirms that the primordial sacrifice—alienation, dismemberment, and reintegration—of Prajāpati or the Purusha is ongoing, continuing, in its sacramental and liturgical re-creation; or else the entire creation would slip directly back into oblivion.[49] The destiny of the whole has ever been what is at stake in the liturgy, the *leit-ourgia*, the work (*ergon*) of the people, the way humans work together with the Divine and the living cosmos. One of the most basic and pervasive functions of ritual (sustaining the *ṛta*, the order of the whole) is well expressed in the Sanskrit *lokasamgraha*: literally "holding (*graha*, to grasp) this place (*loka*, locale) together (*sam*)."[50] More emphatically yet, Ezra Pound's translation from the Tibetan ²Ndaw ¹bpö ritual:

> If we did not perform ²Ndaw ¹bpö
>   nothing is solid
> without ²Mùan̲ ¹bpö
>   no reality.[51]

Considering the machinations of all the presidents, premiers, congresses, and politburos around the world, one cannot help being just a little grateful that there yet remain some such monks and lamas, as well as some people quite ordinary in other respects, who are still holding the world together while the rest of us bumble around making a shambles of it. Indeed, if one required proof of miracles, the very fact that the world still exists 40 years after Hiroshima might be the one datum that clinches the case. It is a statistical improbability, to say the least. Until the H-bomb, it should be recalled, we human beings have not *ever* invented a weapon that we did not eventually get around to using on ourselves. Whatever the reasons for this period of grace may be, it is clear we are living on borrowed time.

Ritual serves as a kind of container for powerful psychic and social energies—dreams, visions, convictions, commitments, enthusiasms, strong loves and hates—which can and do wreak havoc in any society until they are properly channeled. Yet there are no ritual constraints on the escalation of tensions in the nuclear world; the only constant in such a world seems to be the wild acceleration of all social transactions. The net effect is degenerative: an entire society spinning out of control from paroxysm to paroxysm, flying apart like a racing engine without a governor. The very mechanism of our society betrays it; any and every social interaction threatens to hurtle mindlessly beyond all limits. Just so, Gregory Bateson likens the nuclear arms race spawned by the bomb to a terminal, multilevel, bureaucratic addiction:

> The doctrine of deterrence assumes. . .that the networks of interaction between nations are self-corrective and that negative feedback will result from contributing to the armaments race. But the truth is somewhat more complicated. As is commonly the case in biological systems, the short-term deterrent effect is achieved at the expense of long-term cumulative change. The actions which today postpone disaster result in an increase in strength on *both* sides of the competitive system to ensure a greater instability and greater destruction if and when the explosion occurs. It is this fact of cumulative change from one act of threat to the next that gives the system the quality of *addiction*. The addict may think that each "fix" is like the previous fix, and indeed each is alike in staving off the feelings of deprivation. But, in truth, each fix differs from the previous fix in that the thresholds and magnitudes of all the relevant variables have shifted.[52]

In other words, to maintain the feeling of national security, more and bigger "fixes" of weapons are required, and a lethal outcome increasingly likely. Today the U.S. government publicly distrusts its own deterrence policy of mutually assured destruction; what happens when futuristic Star Wars hallucinations no longer give people that momentary feeling of security they seem to require? In our age of unblinking technical and social acceleration, we seem overall to have forgotten the salutary offices of the Sabbath: to pause, relax, take a respite, take stock of ourselves and one another. It is said that anxiety is now the most prevalent mental disorder in this country. Is it any wonder? "Pause a while, and know that I am God," the psalmist once recommended.[53] Lewis Mumford's reflections on the inception of the Sabbath have a critical edge:

> The institution of the Sabbath was, in effect, a way of
> deliberately bringing the megamachine periodically to
> a standstill by cutting off its manpower. Once a week
> the small, intimate basic unit, the family and the
> Synagogue, took over; reasserting. . .the human
> components that the great power complex suppressed.[54]

Of late this deep-seated need for Sabbath is most clearly manifest in the international political arena through that one highly charged word: FREEZE. He who rides the tiger cannot dismount—unless the tiger takes a nap, or loses its appetite for nuclear weapons. Arthur Waskow reminds us why this is so:

> But the human race, in its celebration of secularism,
> has taken no Shabbat for these 400 years. . . .We need
> Shabbat; and Shabbat is the acceptance of a Mystery,
> the celebration of Mystery rather than of Mastery. It is
> the acceptance of the mysterious truth that our own
> Mastery cancels itself out, is most destructive when it
> is most complete.

So the one great word the religious traditions have to
say to the human race and its nuclear arms race is:
MAKE SHABBOS! Pause. Rest. Reflect. Catch your
breath. Meditate. Reevaluate.[55]

Here we must also take into account one of the most
provocative intuitions regarding the function of liturgy.
Liturgical invocation is not magic; it makes no claim to control
the reality it invokes. Prayers for peace carry no guarantees.
Whatever happens in liturgy happens not due to the cause-
and-effect mechanisms detected by conventional empiricism
(mainly the mechanisms of force), but by virtue of the wholly
acausal connecting fibers running like golden threads through
the total fabric of our vital relationships. Through liturgy we
reappropriate our connections with the whole of Life. Through
liturgy, we discover how to dwell within that mysterious
relational integrity by which opposites may attract, and
polarities may interact creatively, instead of merely destroying
one another. Is it not the traditional role of liturgy in the most
farflung cultures to integrate and resolve apparent oppositions
and conflicts into the complementary facets of a more comprehensive understanding? Is this not what is symbolized by the
wonderfully diverse mandalas adorning temples, mosques
and cathedrals alike? Yet outside of the monasteries and
convents, and a handful of recent counterculture experiments
along these lines, western civilization has by and large lost
touch with the many integral visions which once served as its
wellsprings, and which may still serve as mainsprings for
renewal.[56] We have specialized ourselves and forgotten our
responsibility to care for the whole. We have contented
ourselves with the merest fragments of a world, hiding
ourselves away among its broken bits and pieces, which offer
neither refuge nor consistency when the whole is threatened.

By the same token, evidence could easily be adduced to suggest that in most of the traditions which have maintained an integral vision of Life in all its dimensions, there also abides the conviction that this Life is ultimately mystery; that God is essentially unknowable and nonmanipulable; that the Divine, even as it enters and refreshes human life, is sheer spontaneity, unconditional freedom, utter unpredictability, an implosion or explosion of growth from the primordial moment of creation, or else a continuous cycle where frenzies of destruction are followed by triumphal periods of regeneration. How can you know the knower? You can't—it would be merely the known—unless to a certain extent you become the knower. How can you understand God? How can you realize the entire reality? *Tat tvam asi*, as the Upanishads say: that thou art.[57] In a remarkably similar way, the Christian understanding of the human person is capable of discovering the Christic reality that lives and breathes and sees and speaks in and through you. It is nobody else. You are the person, the *thou*, the other pole of the entire reality. Not by having, not by knowing, but by *being*—wholly and uniquely yourself, fully alive, awake, aware, and alert. In such a vision, one does not grasp the whole of reality, but one is so to speak grasped by it, possessed, taken hold of by the whole, by the integrity of the entirety. One does not see this focus—the ultimate concern, the center, the source—one sees *through* the focus, and by this vision the world is made new. The way to the whole is through the center (nexus, nucleus) that is the human person. How else would you get there? According to its etymological roots, the *persona* is not the soul, it is the soul's "mask." But behind the mask there is nothing, or more precisely, *no thing:* not a *what* but a *who.*

And when the poet insists we "should know EVERYTHING . . .at once and always"? Is it not to acknowledge how little

one truly knows, and thus to know that Life is at its core sheer
mystery?—not like a Rubik's cube (to be puzzled over and one
day mastered), but as an ever inexhaustible gift to be celebrated
anew each day. If the nuclear issue—the crux, the heart of the
matter—is human(e) relativity, we have first to recognize that
these relationships reverberate through Heaven and Earth.
The human(e) being finds himself or herself constitutively
related to the Divine and the cosmic, to the atoms and the stars,
the angels and the demons, and to the host of fellow creatures
in between. It is this relational awareness that is most vividly
symbolized by the Christian trinity,[58] the Vedantic *advaita*[59]
(nondualism), and the Buddhist *pratītyasamutpāda*.[60] These
are the living symbols for *that by which the whole coheres*. The
key to the integrity of the whole is neither one nor many. *Neti,
neti*: not this, not that. Trinitarian or nondualistic awareness
breaks through the intellectual impasse of the *hen kai polla*
(one and the many) with which western thinking has been
wrestling for more than two and a half millennia. It is the
great Middle Way between extremes, perhaps never more
rigorously expounded in the West than by the mysterious
"Pseudo-Dionysus," in his *Divine Names* for example:

> Now, the first thing to say is this: that God is the
> Fount of Very Peace and of all Peace, both in general
> and in particular, and that he joins all things together
> in an unity without confusion whereby they are
> inseparably united without any interval between
> them, and at the same time stand unmixed each in its
> own form, not losing their purity through being
> mingled with their opposites nor in any way blunting
> the edge of their clear and distinct individuality. Let
> us, then, consider that one and simple nature of the
> Peaceful Unity which unites all things to Itself, to

themselves and to each other, and preserves all
things, distinct and yet interpenetrating in an
universal cohesion without confusion.[61]

Theology, ontology, metaphysics and today, religious
studies of all sorts, no longer make obnoxious claims to be a
higher kind of knowledge, or queen of the sciences or
whatever, but such disciplines do in fact merit a special
distinction. In our "information age" of mind-numbing data
overload, faced as we are with the fragmentation of all
knowledge and values, with the routinization of life into ever
narrower channels, with the constriction and hyperspecializa-
tion of every human activity, religious studies is most
intimately concerned with the whole of human life—all its
faces, voices, and dimensions—in a constructive way which
cannot be said to characterize any other program of study at
the university. Indeed, much of modern philosophy falls
short precisely here, in its lack of concern for the whole. The
positivistic philosophy which underpins the established
interests of science, government, and the military all too often
uncritically invokes only the reasoning reason as the final
arbiter of what is real and what is not, thereby leaving out
almost everything of any truly human interest. In religious
studies, by contrast, ours is the particularly stringent discipline
of interdisciplinary study per se. The current threat of thermonuclear Armageddon does not require those of us in religious
studies to do something else, or divert our attention elsewhere,
but only to do better what we are already doing, to make
these vital connections more clearly and more trenchantly.
We are the crossroads, the clearinghouse for everything from
the sociology of knowledge to the psychology of human(e)
development, from linguistics and philology to literature,
music, art history, and architecture. I can rarely say this

without getting an argument from the specialists, but I am willing to defend the thesis that religious studies has no subject and no object of its own. Our business may very well be to meddle (discreetly) in everybody else's business.

Under normal circumstances, if circumstances are ever really normal, this would seem to be a quite fortunate dispensation for academic life, since we would have theoretically the freedom to undertake the religious study of just about anything under the sun. In the nuclear age, however, the position of students and scholars of religion can be a particularly excruciating one. Perhaps because it is the whole cloth of our terrestrial, social, and spiritual life that is now threatened, it is in religious studies that we must face directly the questions of how to put it all together again, how to see Life whole again, how to renew the entire creation, how to snatch human life from the jaws of thermonuclear megadeath. There is an almost ascetic catharsis required of western civilization today. In order to transform the lethal structures of our present inhuman(e) condition, we are going to have to learn how to decondition ourselves, to extirpate a great many of the social, cultural, and religious conditionings which have brought us to the brink. This will not be easy; indeed, nothing may be more difficult. But *nothing* is exactly what we may be left with if we do not at least try.

A specific example may be in order here. One such paradigm we have had passed down to us by our forebears is that of the Apocalypse. Now *apocalypsis* has a positive meaning: breaking open (*apo-*) the hidden (*calypsis*), the breakthrough that occurs when opposites turn into one another; not the conflict, but the ensuing enantiadromia. "War is war perverted," says Norman O. Brown; "find the true war."[62]. Unfortunately, the dynamic of Apocalypse has more often been interpreted literally, from one side or the other, than it has been assimilated in its entirety. For a world bristling with more than 50,000 nuclear weapons, it has become a particularly pernicious way to look at human affairs: *us versus them.*

Most apocalyptic scenarios are structurally reducible to a conflict, variously weighted, between an in-group and an out-group, the good guys versus the bad. Usually, as in the Christian Apocalypse or the Jewish Armageddon, the good guys win. Occasionally, though, as in the Zend Avesta or the Norse Ragnarok, the powers of darkness overwhelm the legions of light, or the Giants overthrow the Gods. But apocalyptic eschatologies split the human domain right down the middle. Perhaps as long as we insist on viewing all human relations in terms of us (we are reasonable, rational, and well-intentioned) versus them (they are unreasonable, irrational, and malicious), the human world so constituted will continue to veer precipitously toward cataclysm, and one fine day go over the edge. Certainly such apocalyptic paradigms have been rendered obsolete by the all-inclusive devastation nuclear weapons would wreak, but we still pretend to ourselves that such scenarios are humanly viable. Consider the Middle East: here you have three monotheistic religions, each with its own apocalyptic story and its own justification for "holy" war (the Jewish *herem,* the Islamic *jihad,* and the Christian crusade), each seemingly convinced there will be a final battle where their particular group will triumph, and all of them together making a bloody mess of everybody's lives in the region, each in the name of their God. If we cannot envision alternatives to Apocalypse, then we are by God going to have one.[63] Bear in mind that there are not a few apocalyptically-minded Christian fundamentalists working this very day in U.S. nuclear weapons plants, all of them keenly intent on doing the Lord's work by building bigger and better bombs. Heaven help us!

Another version of the same old thing: our modern-western-Judaeo-Christian-Islamic-Marxist-secular heritage needs revision in very many respects, but this self-congratulatory doctrine of the "chosen people" going to Heaven while the Earth and everybody else on it goes to the devil is sheer nonsense in today's world. Since 1945, wherever we may go, we're all going together. Put negatively, there are no chosen people except those (damned for pride, inevitably) who choose to think themselves saved already. Put positively, we are all God's chosen people. To leave anybody out would presume to put limits on God. It isn't us or them, it's all of us or nobody. To the extent that the western religions have collaborated in building the edifice of history, these religious visions and institutions have also and equally contributed to the cultural, political, and ideological forces today propelling us toward nuclear holocaust. Therefore to the study and the student of religion falls a great part of the task and the responsibility of reviewing, revising, reformulating, rethinking, and renewing the classical sources of meaning and value in our culture. Religion and culture may be distinguished one from the other for any number of reasons, but they cannot be severed. Wherever we find the one, we find the other already present. Studies in religion and culture would be well advised to begin from this integral relationship that the sacred and the secular bear to one another instead of trying to add up a synthetic unity by canvassing the regional ontologies. The whole is surely more than the sum of its parts, and you can scarcely expect to get to the whole merely by adding up the bits and pieces. These provisos in mind, I would go so far as to suggest that it falls to us in religious studies to begin to bring the human family together again—not in a monolithic unity, but in an ever deepening dialogue that takes place always *hic et nunc*, here and now, and never elsewhere. The kingdom of peace is neither above us, sublimely unconcerned with human affairs, nor within us or our group like some kind of private property we have to defend against interlopers: the kingdom is between us, it is in our hands, for better or for worse.

"Where two or more are gathered in my name,"[64] an unexpected and astonishing "increment of association"[65] comes to light, producing "a whole unpredictable from the sum of its components taken separately,"[66]—to cite sources as disparate as theology, economics, and chemistry. Remember the story of "the hundredth monkey," first related by Lyall Watson?[67] Up to a point, Japanese monkeys faced with a new foodstuff learned how to prepare it by imitation; monkey see, monkey do. But at that point, apparently, the added awareness of just one more monkey catalyzes something in them all, and the entire tribe spontaneously learns the new trick of washing the grit off their sweet potatoes. Such things do happen, and not just among monkeys. . . Perhaps the most venerable word (and certainly the most descriptive) in the Christian tradition for this seemingly miraculous redoubled effect of cooperative action is St. Paul's use (in 1 Cor. 3:9, for example) of the rich Greek word *synergasia*: "For we are God's fellow-workers,"[68] or co-workers, or collaborators, that is to say, synergists. *Syn*, together; *ergasia*, working; and the whole— the liturgy, *leitourgia*, this work of the people—demonstrably more than the sum of its parts. R. Buckminster Fuller has recently revived the term, and opened up many physical correlates for this originally theological principle in his *Synergetics*—a design science he defines as "the exploratory strategy of starting with the whole." In the more classical language of theology, one might formulate the basic principle of *synergy* as follows: in any act of salvation or regeneration, God, humans and the entire cosmos are spontaneous cooperators. What this formulation stresses is that the whole, the entire reality, is always implicated in any effort toward salvation, an intuition never more obvious than it is today. In the nuclear age, individual salvation ceases to be the paramount concern, but instead or in addition it is the salvation of this Life here on Earth which rivets our attention. At least we have recognized that the two are intrinsically connected. As J. Schell puts the case, we have not two souls, one to confront the ultimate vacuity of nuclear annihilation and another with which to go about our daily business. The integration begins to take shape, it cannot too often be reiterated, in the crucible of the human person, the living nexus of all these relationships. Our fate is as formidably linked to the fate of the Earth as it is to the destiny of the Divine and the lot of our fellow human beings. In any regeneration, the entire reality—God, human-kind, and universe—must inevitably cooperate. As William Blake once observed, "Everything that lives/lives not alone/ or for itself."

It should by now be clear that one can neither utterly isolate nor totally amalgamate these three irreducible spheres of the real. Reality everywhere displays this pluralistic character, which will not be telescoped or monopolized by any single perspective. Hence the overriding necessity for interdisciplinary research on the frontiers where science, the humanities, and religious studies intersect. Or else the whole thing disintegrates, and we are left with the sciences accelerating into a future nobody wants to inhabit, the humanities nostalgically hankering after a past that cannot be recaptured, and the religions attempting to celebrate a present beyond redemption—a barren, featureless no-man's-land with no familiar landmarks and no guiding lights on the horizon. It is one of the purposes of this book to underscore the myriad alternatives to such a bleak human prospect. The synergy of ways to the center begins with each person, but it does not stop there. Each of us in and through the core of our personal being reaches the highest and deepest poles of the real, each of us contains universes. But it takes two, at the very least, to construct a livable human world. The following chapters will examine some of the ways we might go about it.

The Echo:
An Interdisciplinary
Agenda

And those that create
out of the holocaust
of their own inheritance,
anything more than a
convenient self-made tomb,
shall be known as *survivors*.
Keith Jarrett[69]

Prologue:
The Nuclear Issue
in the Classroom

We have said that the nuclear issue has a human nucleus, meaning that it takes root in that primordial dimension of the human person where the meaning and the mysteries of Life, Death, self, other, change, growth, and so on all coincide: the religious dimension of human life. It is not a monodisciplinary issue, obviously; it is the urgent face of everything that is of concern to us. We earlier emphasized the need for dialogue as the proper way to begin to incorporate these questions of human extinction and nuclear holocaust into the classes we teach and the curricula we are called upon to design. This should take place on at least three levels: dialogue between faculty from the diverse disciplines which must be brought to bear; dialogue between students and faculty in the classroom;

and dialogue between the students themselves, their friends, and families. In point of fact, concern over the nuclear issue in this country and Europe has of late been a truly grassroots movement, beginning with people from every walk of life protesting and crying out in horror, and often only thus filtering into academia from the outside. Peace has lately become the common symbol uniting a variety of activist groups from formerly disparate traditions, cultures, and ideological orientations.

It goes without saying that to incorporate the nuclear issue into courses we are accustomed to teaching in a quite different fashion will not be an easy transition or an automatic procedure. There is the a priori question of one's commitment to studying the issue—no commitment, no study. But even those who are more than willing to devote some portion of their intellectual energies to grappling with this issue will surely run into other stumbling blocks, however strong their commitment. There are always administrative and bureaucratic snags. Moreover, whatever our hearts may try tell us, it is not just the translation of technical jargon that makes us uneasy, and often hugely awkward, in talking about megatonnage, throw-weight, counterforce and countervalue strikes, and so forth. This reluctance, even awkwardness, in bandying about the lingo of war and weaponry is by no means a bad thing in itself. It marks the threshold between the creative scholarly milieux we should prefer to inhabit and the destructive milieu of the war machine our civilization has become. It is not only a depressing topic, it is as Lewis Mumford has pointed out since August of 1945, literally a demoralizing research;[70] it is research into the de-moralization of the cultural integrity of an entire civilization, forcing itself into the mold of a monster with two heads, the Russian and the American, each of which appears determined either to rule over or to destroy their common body, the Earth, and all her children. All the more reason for such research to go on in our universities, re-search in the original sense: "searching again" the entire range of social paroxysms we call the nuclear issue, never contenting ourselves with partial answers to this all-inclusive human question.

Since we have so freely filled previous pages elucidating some of the neglected aspects of this hydra-headed question, it would not be amiss at this point to acknowledge that, yes, the existence of 50,000 thermonuclear bombs is also part of the nuclear issue and must be dealt with on its own terms. Until our civilization learns to see all these fancy and costly missiles, along with their attendant hardware, as so many small mountains of nuclear waste (which we don't know how to get rid of either), it will be necessary to study the development and deployment of this infernal weaponry, as well as what the effects of using it might be. These systems must be closely monitored by all of us. As we shall see, there is really nobody else in charge, no firm hand on the nuclear tiller of today's wayward ships of state.

The Past
(Religion and Science)

Nuclear weapons exist. The know-how to make more of them also exists. The risk of nuclear holocaust can and should be reduced nearly to zero by every conceivable means, but (as we noted earlier about the accompanying possibility of extinction) the proximity of "the fire next time" will endure precisely as long as the human species endures. The human mind has no reverse gear, and it seems that the brakes are bad as well. When Alfred Nobel invented dynamite in 1866, he supposed that with such a terrible weapon in the world's arsenals nobody could ever countenance waging war again. His contemporary, Dr. Josiah Gatling, invented the machine gun during the Civil War with an apparently similar humanitarian motive. Since the dawn of the nuclear era, this kind of innocence is lost forever. Perhaps, centuries earlier, the Second Lateran Council of 1139 was more modern than medieval in its outlook when it banned the use of the crossbow on the grounds that human life should not be taken "automatically" by a mechanical device.

Although they are many orders of magnitude more destructive than TNT, gatling guns, and crossbows, nuclear weapons did not just fall from the sky into our midst, nor were they disgorged up from Hell. These instruments of genocide are the direct result of a series of distinctly human decisions, discoveries, desires, and delusions occurring in Europe and America during the first half of the 20th Century. The quantum physics involved in the truly extraordinary feat of building the first atomic bombs is fascinating in its own right, of course, but it bears on our concerns in religious studies in a very particular way. The evolution of nuclear physics over the past 90 years or so can be seen as a nearly perfect case history of the catastrophic failure to link theory to practice. It is an extremely clear lesson in (how not to go about) human(e) relativity. It must not escape our notice that much of modern physics has very quickly evolved in this century from a recondite abstract science into a government-sponsored weapons-procurement program. Certainly there is still a little disinterested research going on, withstanding the latest flood of glittering Star Wars grant money and inducements, but what is indeed disappearing is the very presumption of a purely objective science, free from any ideological bias or moral intentionality. As we shall see, the bomb began to awaken the long-slumbering conscience of the scientists. We are all a little more wary now. We know these things can get out of hand and backfire grievously. We are more than ever aware of environmental repercussions. And, happily, we are learning how to raise the more general and perennial questions of exactly what sorts of relationships really do or ought to pertain between human beings and the biosphere we share. Some highlights of the crucial scientific developments at issue here are therefore in order. The following is not an exhaustive list of topics, or an alternative history of nuclear physics in the 20th century. It is merely a sampling of the kinds of items that ought to appear on the agenda for further research:

- 1896 — It should not be forgotten that the *atomos*, the indestructible monad at the basis of western metaphysics, began to reveal its innate "self-destructive" tendencies when the French chemist A. H. Becquerel managed to expose a photographic plate with the radioactive emissions of uranium. From this moment onward, the physical world begins to lose its solidity—an event, as R. Panikkar remarks, of enormous anthropological consequences with which we have not fully reckoned to this day.[71]

- 1897 — It should not be forgotten that J. J. Thompson's discovery of the electron soon confirmed that the once supposedly indivisible atom indeed had constituent parts. The electron was of course only the first of many subatomic components to come to light, but already the suspicion could be justified that there is no sub-stance standing firmly under the world of matter—a discovery which at a stroke renders obsolete all substance metaphysics and any form of materialism (dialectical or otherwise).

- 1906 — $E=mc^2$. It should not be forgotten that, among other startling illuminations, Einstein's Special Theory of Relativity postulated the convertibility of energy and matter. Matter is mainly a matter of perspective.[72] In theory, it was already clear that tremendous energy was locked into the strong binding forces of the atom, although it took nearly 35 years to work out a practical way to release the energy by bombarding nuclei with electrically neutral particles. But the Theory of Relativity had a more subtle and more penetrating effect, too often overlooked: it threw us all into a precarious web of interdependence where not only energy and matter, but nature and culture as well, are inextricably intertwined.

• 1924 — It should not be forgotten that Niels Bohr's principle of complementarity did not resolve the much-disputed wave/particle problem, but it did add to the mathematical discussion a very human truth: you see what you look for, and sadly, often *only* what you look for. "A complete elucidation of one and the same object," wrote Bohr, "may require diverse points of view which defy a unique description."[73] If you test for wave phenomena (interference patterns, etc.), that is what your test results will confirm. If you test for particulates, then even light (e.g., aimed through a pinhole), seems to strike a surface in discrete quanta. Each of the two then-current theories of atomic structure—as a matter/wave continuum (Schrödinger/di Broglie) or as discontinuous particulates (Heisenberg)—could mathematically account for most of the available data,[74] but neither theory could account for the existence of the other, equally plausible interpretation of the same data. Bohr summarized the dilemma neatly: "The opposite of a trivial truth is plainly false. The opposite of a great truth is also true." Now once upon a time, Aristotle had held as self-evident the principle that "the same thing cannot at one and the same time be and not be, or have any other similar set of opposites,"[75] and this principle of non-contradiction (the so-called excluded middle, the dialectical either/or) has in fact served for 2,500 years as the predominant pattern of intelligibility in the West:

A ≠ B, a chair is not a table, I am not you, we are not
them, a particle is not a wave, and so on.[76] Things are
defined or individuated in terms of their differences
from other things. But now quite suddenly, in the
microcosm of the atom and in the human mezzocosm
as well, we find that any true statement must in some
manner and to some extent include its opposite, and
will therefore always display the character of a qualified
tautology.[77] Bohr and Einstein's later epistemological
dialogues on these issues testify in another way to this
very human aspect of nuclear physics, namely that you
have to take seriously the other fellow's opinion,
because he just might be right from his perspective, and
you might well be wrong or partial or one-sided unless
you learn to take that perspective into account.

• 1927 — It should not be forgotten that Werner
Heisenberg's principle of indeterminacy (sometimes
called uncertainty) soon clinched the case: human
knowledge is always provisional. "The only way for the
observer, including his equipment, to remain uninvolved
with what he observes would be for him to observe
nothing at all."[78] Heisenberg's principle, reduced to its
barest outline, asserts that you cannot at the same time
determine the exact position and exact velocity of an
atomic "wavicle." Why not? Because to gauge position,
the observer has to move, to take a fix as as a surveyor
does from at least two angles. (In the subatomic realm,
of course, the role of observer is largely played by the
observing apparatus, e.g., by a stream of charged
particles which interacts with whatever is observed.) To
measure velocity, on the other hand, the observer or his
equipment must take up a fixed position. You can't have
it both ways at once. The popular philosopher of science

Jacob Bronowski once proposed that Heisenberg's so-
called uncertainty principle be renamed the "principle
of tolerance." This is surely more accurate on several
levels: within certain limits, certain tolerances, we
human observers know what we know tolerably well.
But we must always be open to—that is to say tolerant
of—anomalies and discrepancies in our concepts and
descriptions because we can never attain absolute
certainty in our descriptions of the physical world. So
much for objective facts in science, which turn out
always to be facts for human subjects and therefore
a palimpsest of interpretation(s) at every epistemological
and ontological level.[79] Heisenberg himself would say
that we no longer have a science of nature, we have
a science of the mind's knowledge of nature.[80]

- It should not be forgotten that in the early 1930s,
while the atomic scientists were busy refining this
principle of tolerance with enormous mathematical
subtlety and precision, the monstrous intolerance of the
Nazi era with all its massive certainty and efficient
brutality was gathering steam well beyond the boiling
point all around them. The "Jewish science" of nuclear
physics was quickly disenfranchised, and the exodus of
scientists from Göttingen and Heidelberg to England
and America began in earnest. The scientists found
themselves rapidly and thoroughly politicized, for
obvious reasons.[81] There were hidden undercurrents as
well. Rutherford's well-equipped laboratory in England,
for example, was the first atomic physics facility to
receive government funding, long before the Manhattan
Project got underway in the States. Historian Robert
Jungk remarks that at the time, the physicists seemed
not to suspect (or care?) that one day he who pays the

piper is going to want to call the tune.[82] Today there can be no doubt who is calling the nuclear tune.

• 1935—Generally speaking, it has been quite forgotten that as early as October 1933, Leo Szilard had surmised that a nuclear chain reaction could take place on a large scale, and by 1935 he was approaching a number of his colleagues to ask them, as Jungk relates, "whether it would not be advisable, in view of the possibly momentous and perhaps even dangerous consequences of their present studies, to refrain, at least for the time being, from publishing any future results of their investigations. His suggestion was for the most part repudiated."[83] The physicists passed their first crisis of conscience with barely a qualm.

• But no one forgets that it was in late December 1938, that atomic fission was first achieved by Hahn and Strassmann—inside Nazi Germany. The race to develop the atomic bomb dates from this event. It was, however, a one-sided race. Physicists Hahn and Heisenberg and von Weizsächer met secretly after the successful fission experiment and agreed to prevent (as only this group could) the development of an atomic weapon for Hitler's Reich.[84] The strategy appears to have been for Heisenberg and the others to retain control of atomic research inside Germany. To do so required at least the appearance of collaboration with the Reich, a course of action which naturally made their former colleagues very nervous indeed. Heisenberg later maintained that "under a dictatorship active resistance can only be practiced by those who pretend to collaborate with the regime."[85] As Jungk tells the story, however, even his friend Bohr remained skeptical when Heisenberg cryptically attempted to signal his true intentions in late 1941; so skeptical that he began immediately thereafter to encourage the Anglo-American military authorities in their efforts to anticipate the construction of an atom bomb by the Nazis. In retrospect, it is only fair to concede that all this paranoia must have seemed amply justified at the time. Even the cautious Szilard was alarmed enough to convince Einstein to sign the famous letter to President Roosevelt (2 August 1939), which in turn convinced the president to authorize creation of the Manhattan Project.[86] Einstein long afterwards termed signing that letter the greatest mistake of his life, but the deed was done.

• It should never be forgotten, however, that the rationale of beating Germany to the bomb disappeared in early 1945 when physicist Samuel Goudsmit and his ALSOS mission inside the conquered Germany determined that the Germans were not even close to assembling a bomb,[87] presumably due to magnificent and underestimated stalling by Heisenberg et al. A few scientists at Los Alamos raised the obvious question: "Why finish building the bomb if Germany doesn't have one?" But Oppenheimer reputedly dismissed the suggestion out of hand. Indeed, as Jungk relates, "Never was the pace at Los Alamos fiercer than after the capitulation of the Third Reich."[88] By 1945, the momentum of the project seemed irreversible. The physicists had easily passed their second possible crisis of conscience.

Here then is the tragic juncture where practical accommodations and military applications on the part of the atomic scientists began to diverge radically from the theory and all the grand implications of complementarity and indeterminacy. Here the principle of tolerance, under any other name, did not reign supreme. The physicists and engineers gathered under J. Robert Oppenheimer's direction at Los Alamos in some part to refine their knowledge of the mysterious inner world of the atom, of course, but mainly in order to contribute to the war effort in a dramatic way, to prevail against the great evil, to fight fire with fire, literally, and on a massive scale. But they too were seeing only what they looked for, from a hugely provisional perspective in the midst of the stymied communications of a world at war. They were crusaders: God, truth, the good, and the forces of civilization were on their side. Hitler looked like evil incarnate in a man, "the one out of Hell." What the scientists neglected to take into their computations were the very demonic dynamisms taking shape within them and between them, dynamisms which would soon bring an even greater and more permanent evil into the world. In so many words, they knew better—an idiot could guess that an "ultimate" weapon invented in wartime would be used as a weapon of war—but they did it anyway. They built the damned thing. It is hardly necessary to offer retroactive moral condemnations of the physicists' collaboration in the building of the bomb; they've been sorely vexing their own consciences ever since. Oppenheimer himself used some of the strongest language. As he put it in the summer of 1956, "We did the devil's work."[89]

So far we have begun to glimpse how physics, moral philosophy, cosmology, and cross-cultural studies have crisscrossed each other in even this short list of historical developments. One ought also to remark a series of what seem in retrospect to be rather startling political decisions authorizing, accompanying, and reinforcing the physical development and military deployment of our tiny wartime atomic arsenal:

- It should not be forgotten that on 6 December 1941, Roosevelt signed the presidential order authorizing the Manhattan Project—the day *before* Pearl Harbor was bombed, and thus before the United States was formally at war with anybody. This Manhattan Project was at the time the single most enormous and most expensive scientific undertaking in the history of the human species, though minute by the bloated standards of today's "defense" budget. In short, we did not develop the atomic bomb by accident.[90]

- It should not be forgotten that the firebombings of Dresden and Tokyo late in the war were deliberately and precisely engineered to create self-fueling firestorms. Months of strategic planning were required for saturation bombing to produce an inferno equivalent to that generated by an atomic detonation. As mentioned earlier, it is at this juncture that slaughter of noncombatants becomes explicit Allied policy, thereby setting the stage for immediate acceptance of the atomic bomb as merely a more efficient weapon for the same purpose: mass murder. It is not often enough noted that approximately half a million people were killed in these B-29 fire-bombings, more than double the number of deaths in the atomic bombings of Hiroshima and Nagasaki combined.[91]

Several of the dates that follow will probably never be forgotten, or their consequences erased from human memory:

- At 5:30 a.m., 16 July 1945, in the desert called *Jornada del Muerto* (Journey of the Dead) near Alamogordo, New Mexico, the Trinity Test of the first atomic "Gadget," as it was called, is a stirring success. Why "trinity" test? It is said that this seemingly blasphemous code name was coined by Oppenheimer himself, who happened to be reading John Donne's Fifth *Holy Sonnet*: "Batter my heart, three person'd God." In theological retrospect, the name is astonishingly accurate: the Trinity Test may indeed represent the most direct challenge to the trinitarian constitution of the real that humankind has ever experienced. The test of the trinity is the test of that principle by which the whole of reality coheres. Some authors, Jim Garrison and Arthur Koestler among them, date the nuclear age from the Hiroshima bombing. I would however opt for the date of the Trinity Test as the real beginning of the end: after 16 July 1945, there is no turning back. Upon witnessing the explosion, Oppenheimer significantly either misquotes or retranslates Book XI of the *Bhagavad Gita*: "I am become Death, the shatterer of worlds."[92] He later put it that "the day after Trinity" may have been the last moment when control of atomic weapons actually remained in the hands of the scientists.[93] From here on, the military strategists take over.

• There was a choice between using the weapons on human targets and staging a demonstration blast—in Tokyo harbor, for example. But the so-called Franck Report of June 1945, which called for such a demonstration and outlined with prescient accuracy the disastrous effects of a runaway atomic arms race, was belittled and rejected by Oppenheimer (what if the bomb was a dud?, he asked) and ignored by the majority of the physicists in their final recommendations to the military.[94] Conscience once more gave way to the press of wartime circumstance. Oppenheimer later summed up the stark dilemma of the atomic scientist, a dilemma to which he at least as much as any other single scientist had contributed: "In some sort of crude sense which no vulgarity, no humor, no overstatement can quite extinguish, the physicists have known sin, and this is a knowledge they cannot lose.[95]

• 17 July-2 August 1945, the Potsdam Conference brings together the Allied leaders to implement the Yalta agreement and divide the spoils of the victory in Europe. As the conference opened, President Truman was informed by General Groves of the success of the Trinity Test, and was soon calling for Japan to surrender unconditionally—or else suffer "prompt and utter destruction," by means never specified. Nowadays it is often maintained that the use of atomic weapons against Japan saved a million American lives that would have been lost in a prolonged land invasion. But there is convincing evidence that Japan was already seeking to surrender through diplomatic channels, and that invasion of the Japanese mainland would not in fact have been necessary. Apparently the OSS was aware that Prime Minister Tojo had wired his Russian Ambassador Sato in mid-July: "Japan is defeated. We must face that fact and act accordingly." But the Russians made light of Japanese attempts to negotiate a conditional surrender (the condition, eventually met, was that they be allowed to retain their emperor); and the Americans ignored them altogether. Calculating as it might sound, Truman's main interest in these final days of the war may well have been to demonstrate America's newfound atomic capability in order to secure political preeminence in the postwar world.[96]

• August 1945, the atomic bombing of Hiroshima: 80,000 dead on impact, another 28,000 eventual casualties from radiation effects. Some 90 percent of the physical structures of the city were leveled by the blast. Hiroshima had

deliberately been spared conventional bombing so that the
effects of the atomic weapon on both military and civilian
targets could more clearly be gauged. (It has since come to
light that the U.S. Army was at the time aware of several
American POWs, their jail cells less than a kilometer from
ground zero, who were also incinerated in the indiscriminate atomic explosion.[97] Acceptable losses?)

• 9 August 1945, the atomic bombing of Nagasaki: 75,000
casualties. The survivors—called *hibakusha*—become
pariahs, and the physical, psychological, and social
scars of the atom bomb on many levels of Japanese society
are still plainly evident today.[98]

• As an important sidelight, it should not be forgotten
that the first real computers, the vacuum-tube ENIAC
and MANIAC designed by mathematician John von
Neumann at Los Alamos, were developed for the sole
purpose of quickly performing the massive calculations
necessary to produce thermonuclear weapons.[99] It has
been said that without these calculating machines, there
would not have been enough man-years left in this
century to complete the computations. The computer has
been intimately allied with nuclear weapons ever since.
Computer control and guidance systems today govern
the entire strategic arsenals of the superpowers. They are
notoriously unreliable. Every time lightning strikes the
municipal power grid in Colorado Springs, Colorado,
for instance, the North American Air Defense Command
(NORAD) computers there register a Soviet missile strike.

• The history of the subsequent arms buildup has been
traced in any number of serviceable works.[100] The
Russians exploded their first fission device toward the
end of August 1949. It should not be forgotten that the

ensuing arms race was predictable, down to the last damning detail, from the very beginning. The Franck report just mentioned is one early example of such foresight, and Lewis Mumford's several essays on the topic in the late forties are even more proleptic: "Gentlemen, you are mad. . ."[101] It cannot be argued that nobody guessed what would happen if the United States tried to keep its atomic secrets to itself. It is equally important to give due weight to the fact that for the first 30 years, the U.S. was always the first to escalate the quality and quantity of nuclear weapons.

• On Halloween night 1952, the first "Super" or hydrogen bomb, nicknamed Mike, was exploded on the islet of Elugelab in the South Pacific. If the name sounds unfamiliar, this is because the little island of Elugelab no longer exists. It disappeared under the fireball of this first man-made star, leaving only a crater a mile long and 175 feet deep dug into the Pacific basin.[102] The former inhabitants of the surrounding Eniwetok Atoll (one of the Marshall Islands) have recently been permitted by the U.S. government to return to some of the less contaminated islands, radioactive coconuts notwithstanding.

• On 8 August 1953, Georgi Malenkov, Stalin's successor, announced to the world that "the United States no longer has a monopoly on the production of the hydrogen bomb."[103] Atmospheric traces of radiation confirmed not only that the Russians had detonated an H-bomb, but also that it was a so-called dry bomb—lighter and more compact than the American prototype, and deliverable by air. The arms race was on in earnest.

• Finally, it must not be forgotten that there is a patchy, flickering quality to the way the nuclear arms race has been perceived by the American public, who did, after all, pay the bills. This is partly due to a lack of facts— information not declassified for decades, and often ambiguous even when it finally comes into the public domain. But it is also partly due to the strange psychosocial effects of such ultimate weapons. People avoid the issue, and refuse to look at the facts even when these are presented directly, preferring instead to leave the entire issue in the hands of experts whose very careers often depend on obfuscating such things. Example: remember the "missile gap" during the Cuban crisis in 1961? President Kennedy ordered that the United States build its first 1,000 nuclear-tipped ICBMs to counter a supposed buildup of similar missiles by the Russians. Daniel Ellsberg authored the strategic defense plan for the deployment of these new weapons under the supervision of Robert MacNamara, then Secretary of Defense. And how many missiles did the Russians actually have at the time? Four. Years later, Ellsberg discovered to his dismay that his nuclear defense plan had in fact been meant to serve as a first-strike plan, and as such was duly adopted by the Pentagon.[104]

Indeed, the policy of deterrence has mainly turned into an excuse for procuring more weapons. When the momentum of weapons production flagged in the mid-seventies, new "limited" nuclear war scenarios were duly manufactured which in turn required new generations of weapons. If it's no longer the "missile gap," then it's the "bomber gap" or the "anti-ballistic missile gap" or the futuristic "window of vulnerability" profferred by the Reagan administration during its first term. Now it's Star Wars, the so-called Strategic Defense Initiative. Any excuse will do, so long as the public can be cajoled into funding new military procurement. This is big business;

paranoia can be profitable, at least in the short run.

The American public has always perceived itself to be ordering up new generations of nuclear weapons in self-defense, and so of course have the Russians. But today, when the failure of a single 43-cent silicon chip can put the Strategic Air Command on red alert and bring this nation within six minutes of initiating an unprovoked nuclear strike, it is imperative that we try to see clearly to the roots of this lethal paranoia. Because the over-armed superpowers cannot engage one another directly, a constant series of conventional military engagements have taken place since the fifties through surrogates and proxies. After Korea and Vietnam, one would think we should know all too well the price in lives and dollars and lost integrity that this kind of "non-nuclear" cold warring exacts, while at the same time bringing us ever closer to the ultimate conflagration. We are under the gun. The aberrant human relationships spawned by the advent of nuclear weapons on the world scene have crystallized into one of the most anguishing questions our species has ever had to ask itself: can we survive our own paranoia? If we human beings cannot cope constructively with the diversity of human values, cultures, religions, economies, and political systems, then the superpowers will sooner or later run out of surrogate victims and annihilate themselves, and most of the rest of the human family in the bargain. Any real alternative to the present, increasingly unstable, and explosive state of mutually paranoid projections is going to have to begin with constructive dialogue between the paranoid parties, now postponed for over four decades. What passes for arms negotiations in the current international political arena will amount only to window dressing or public relations until the underlying paranoia can be eradicated. In the early days of the nuclear era, C.G. Jung put the case somewhat as follows:

Since it is universally believed that man *is* merely
what his consciousness knows of itself, he regards
himself as harmless and so adds stupidity to iniquity.
He does not deny that terrible things have happened
and still go on happening, but it is always "the others"
who do them. . . . No one will maintain that the atomic
physicists are a pack of criminals because it is to their
efforts that we owe that peculiar flower of ingenuity,
the hydrogen bomb. . . . But even though the first step
along the road may be the outcome of a conscious
decision, here, as everywhere, the spontaneous idea—
the hunch or intuition—plays an important part. In
other words, the unconscious collaborates and often
makes decisive contributions. . . . If it puts a weapon in
your hand, it is aiming at some kind of violence. . . . In
theory, it lies within he power of reason to desist from
experiments of such hellish scope as nuclear fission if
only because of their dangerousness. But fear of the
evil which one does not see in one's own bosom but
always in somebody else's checks reason every time,
although one knows that the use of this weapon
means the certain end of our present human world.[105]

As for the scientific innovations that have thrown us willy-
nilly out of the frying pan of World War II and into the chill
fires of the cold war, we should above all note that there has
been a powerful shift of emphasis in science as it is practiced
since the war. Indeed, the most significant transformation of
scientific method often goes unnoticed when one reviews the
compendium of remarkable discoveries about the physical
structures of matter and energy on which the scientific establishment quite rightly prides itself. The really drastic change here
is the way in which science almost everyhwere has become

centrally administered and systematically applied science,
that is, technology. Technology is not just a handful of tools
anymore, it is a carefully managed system that controls research
and attempts to direct research into predetermined channels.
Even to qualify for funding, most research projects in America
these days must justify their existence by anticipating the
almost instantaneous application of whatever is uncovered,
all too often in the highly lucrative form of weaponry. Over
half of the scientists and engineers in the United States are
currently employed directly or indirectly by the defense
establishment. Too often, such research turns inexorably into
"development" under the terrifying rubric, "If it can be done,
it must be done." The upshot is that technical advances in
weapons systems now dictate policy on a global scale. In
weapons research above all, the assumption is that if a new
weapon can be produced, it must be produced—immediately,
before somebody else gets the jump on us, a shameless inversion
of the Golden Rule: do it unto others before they do it unto
you. As a recent United Nations study emphasizes,[106] it is this
runaway technological momentum, this demonic automatism
of the arms race, that turns all the political and legal avenues
of redress into dead ends. The crucial problem here seems to
be a failure to distinguish between two very different human
attitudes toward the scientific endeavor: on the one hand,
a contemplative attitude (useless, but fruitful) which wants to
know how things really are, and on the other hand a single-
mindedly utilitarian attitude (obviously useful, but also
dangerous) which wants to control and manipulate things.[107]
Ironically, it is the contemplative attitude of basic research
that generally produces the great breakthroughs and original
discoveries; the utilitarian attitude mainly extrapolates from
what is already known.

However this may be, there is no abstraction "science"

existing in some conceptual vacuum. The contemporary
scientific enterprise is not a monolithic edifice, but a constantly
revised set of only partially overlapping theories, hypotheses,
models, and metaphors.[108] The awakened ecological sensitivities
of the past couple of decades have put us on a new footing
from which to evaluate the attitudes and assumptions which
have so long dominated human intercourse with the natural
world, and which have impelled us with seemingly irresistible
momentum into the contemporary nuclear predicament.
What is happening today could be formulated in classical
theological terms as the encounter of the traditionally transcendent God (the Father) with the newly re-emerging
Goddess or Divinity (the Mother) sensed and intuited in the
rhythms, dynamisms, and patterned integrities of the natural
world. Is it to be another clash of superpowers, or a mutual
fecundation?—the *hieros gamos* or sacred marriage of Heaven
and Earth, sacred and secular, God and the world, so long
ignored and practically annulled by the regnant machine age.
Between these two poles, a new cosmology is being forged.
We need to put the world together again in a physical, as well
as in an intellectual and moral sense. We have to learn to see
Life whole. Thomas Berry has made a start with his stunning
attempt to tell this "whole story" of creation from a Teilhardian
point of view which is both scientifically credible and
religiously viable:

> The Story of the Universe is the story of the emergence
> of a galactic system in which each new level of being
> emerges through the urgency of self-transcendence.
> Hydrogen is the presence of some millions of degrees
> of heat emerges into helium. After the stars take shape as
> oceans of fire in the heavens they go through a sequence
> of transformations. Some eventually explode into the

stardust out of which the solar system and the earth take shape. Earth gives unique expression of itself in its rock and crystalline structures and in the variety and splendor of living forms until man appears as the moment in which the unfolding universe becomes conscious of itself. Man emerges not only as an earthling but also as a worldling. He bears the universe in his being as the universe bears him in its being. The two have a total presence to each other.[109]

It was for just such a reunion of scientific precision and religious experience that the often caustic theoretical physicist Wolfgang Pauli—one of the few major physicists, by the way, who refused to work on the atomic bomb project from the outset— appealed in a 1955 lecture on "Science and Western Thought," an appeal that scientists and religious thinkers alike are finally beginning to understand and to heed:

Since the seventeenth century the activities of the human spirit have been strictly classified in separate compartments. But in my view the attempt to eliminate such distinctions by a combination of rational under-standing and the mystical experience of unity obeys the explicit or implicit imperative of our own contemporary age.[110]

Here we can only touch upon the hallowed belief, already ancient by the time it receives the endorsement of the Pytha-goreans, that the world is a living organism. The closest modern correlate may be Lewis Thomas's intuition of the Earth as a single living cell, or else possibly the grand self-regulating system of James Lovelock's "Gaia Hypothesis."[111] As Plato presented the birth and shaping of the world in his *Timaeus*, "We may say that the world came into being—a living creature truly endowed with soul and intelligence by the providence of God...Wherefore he made the world in the form of a globe, round as from a lathe, having its extremes in every direction equidistant from the center."[112] This *anima mundi*—soul of the world—can be expressed most simply as the understanding that the cosmos is alive, animate (that is, "en-souled"), that it has its own integrity and contains its own principles of movement, growth, and regeneration. Through the so-called Neoplatonic tradition, this ancient idea is equally familiar to classical, medieval, and renaissance thinkers.[113] The pitfall here is of course pantheism. To totally identify the world soul with God or the holy spirit would not only abridge the transcendence of God, it would trap human beings in a deterministic world pre-programmed from the start. This proviso in mind, R. Panikkar observes:

> Philosophers and theologians alike, however, took care to distinguish the formal or material principle of the world from God, lest he be degraded to a merely intramundane reality; nevertheless; there was a consensus that a certain unifying principle inherent to the universe made it a unity and, in a certain sense, a worthy partner for Man....And yet this notion does not destroy the hierarchical conception of the universe. On the contrary, it reinforces the hierarchy by locating Man as well as all other superior and inferior beings in their proper places.[114]

I have elsewhere contended that modern physical and biological science is slowly but surely rediscovering the geometrical matrix of the *anima mundi*, but it is equally clear that the pertinent discoveries have by and large occured within contexts so specialized that this renewed vision of the world as a whole—intact, coherent, and abundantly alive— has not yet been fully spelled out. One of the earliest attempts to do so in relatively modern scientific categories was the now-classic *Holism and Evolution* (1926), by the renowned polymath Jan Christian Smuts, who explicated his coinage of the term "holism" in an extraordinarily powerful way:

> The creation of wholes, and ever more highly organized wholes, and of wholeness generally as a characteristic of existence, is an inherent character of the universe. There is not a mere vague indefinite creative energy at work in the world. This energy or tendency has specific characters, the most fundamental of which is whole-making. And the progressive development of the resulting wholes at all stages—from the most inchoate, imperfect, inorganic whole to the most highly developed and organized—is what we call Evolution. The whole-making, holistic tendency, or Holism, operating in and through particular wholes, is seen at all stages of existence, and is by no means confined to the biological domain to which science has hitherto restricted it. With its roots in the inorganic, this universal tendency attains clear expression in the organic biological world, and reaches its highest expressions and results on the mental and spiritual planes of existence. Wholes of various grades are the real units of Nature. Wholeness is the most characteristic expression of the nature of the universe in its forward movement in time.[115]

Arthur Koestler is the latest exponent of Smuts' holism, and has refined it by positing what he calls a "holarchy"—a hierarchy of wholes open-ended in both the microcosmic and macro-cosmic directions, which suggests again the three worlds of traditional religions, with the human domain located midway as a kind of mezzocosm constitutively related to the whole(s) at every other level. Koestler emphasizes that scientific method has learned how to break up nature's wholes (atomic and cellular nuclei, for example) but is only on the verge of learning to detect the integrative principle which holds these realities together. To illustrate this non-causal cohering principle, he assembles a formidable battery of evidence and authorities:

> In the present theory, the "order from disorder" principle is represented by the integrative tendency. We have seen that this principle can be traced all the way back to the Pythagoreans. After its temporary eclipse during the reign of reductionist orthodoxies in physics and biology, it is once more gaining ascendancy in more sophisticated versions. I have mentioned the related concepts of Schrödinger's negentropy, Szent Györgyi's syntropy, Bergson's *elan vital*, etc.; one might add to the list the German biologist Woltereck who coined the term "anamorphosis". . . for Nature's tendency to create new forms of life, and also L.L. Whyte's "morphic principle," or "the fundamental principle of the development of pattern." What all these theories have in common is that they regard the morphic, or formative, or syntropic tendency, Nature's striving to create order out of disorder, cosmos out of chaos, as ultimate and irreducible principles beyond causation.[116]

Along these lines, R. Buckminster Fuller has in his two *Synergetics* volumes[117] articulated what he takes to be "nature's coordinate system" in very precise terms as a spectrum of polyhedral forms, reminiscent of but not identical to those in Plato's *Timaeus,* which so to speak govern the formation of these wholes at every level from the nuclear and molecular (physics and chemistry), through the cellular and organic (biology and medicine), up to and including the balances and reciprocities of human(e) relations between people and peoples (ethics, justice, etc.).

It is perhaps sufficient here if we glimpse the gamut of efforts in this maverick science (not yet quite accepted by the mainstream scientific establishment, but significant nonetheless), which all tend toward an enunciation of the "syntropic" principles by which the whole cosmos coheres.[118] Such approaches imply a considerable realignment of educational priorities. Instead of emphasizing only extrinsic causes, where forces from the outside act upon objects, we should find ourselves emphasizing the intrinsic connections, where each being discovers its intimate relations with all others. As the great biologist Gregory Bateson used to insist, in this nuclear age of rampant disintegration and disorientation, what matters most in education is the ongoing search for "the pattern which connects":

> *Break the pattern which connects the items of learning and you necessarily destroy all quality.*

> *The pattern which connects.* Why do schools teach almost nothing of the pattern which connects? Is it that teachers know that they carry the kiss of death which will turn to tastelessness whatever they touch and therefore they are wisely unwilling to teach anything of real-life importance?

> Or is it that they carry the kiss of death *because* they dare not teach anything of real-life importance? What is wrong withg them?

> What pattern connects the crab to the lobster and the orchid to the primrose and all four of them to me? And me to you? What is the pattern which connects all the living creatures?[119]

The Future
(Religion and Literature)

If the history of the nuclear predicament seems strewn
with errors and oversights and corpses, the future it implies
seems literally to be collapsing before our eyes. With no apol-
ogies to Alvin Toffler, I have elsewhere called this peculiar
state of affairs "No-Future-Shock."[120] Children today grow up
with the not unreasonable conviction that they will die in
a nuclear holocaust. And obviously, if the bombs are ever used,
there will indeed be no future. But this ultimate horror can
come at any moment. Therefore an entire generation has come
of age ever uncertain, always uneasy, and trepedatious about
committing themselves to anything which stretches toward
that grey area of might-not-be. No exit: the feeling is that of

being surrounded on every side by dead ends and chasms, fogbound in a maze. There is no safety, and nowhere left to run. Any step could be our last. In the past we made grievous mistakes, granted; but what have we done to the future?

We have shattered the looking glass of the future, and now stare with unseeing eyes into the void, into our own non-being. We have forgotten how to look forward—to anything. Whether consciously or unconsciously, every human being who has come to adulthood during the past quarter century has been challenged to face head-on the possibility of anthropocide, the Death of humankind, and terracide, the Death of the Earth. Whether we know it or not, we have all had to come to terms in one way or another with the effects of nuclear weapons, whether they are ever used or not, on the body and soul of humankind. These effects are manifold and probably incalculable in all the most crucial respects, but they come to a head in this issue of human extinction. Nobody, as Jonathan Schell pointed out, will ever experience extinction.[121] We have therefore no inbuilt instinctual or cultural or institutional safeguards against species-specific suicide. A chilling thought, to which Schell adds another: we have only one Earth. We cannot stage a nuclear holocaust just to see what will happen. I would maintain that if we are not in some measure able to experience this megadeath *now*—symbolically, metaphorically and, indeed, liturgically—then we may very well doom ourselves and our species to oblivion; physically, literally, and irremediably, with no hope of resurrection. This symbolic death will probably mean the death of some of our too-precious identities (national, cultural, ideological, and personal) but it may be the only reliable source of renewal, for each of us and all. The symbolic death must be real for us, or else the renewal will come to no more than wishful thinking. But how to do it?

The recent past has very nearly murdered the future, and has left the post-Hiroshima generation staggering under the shadow of imminent annihilation. This "future-loss" has in its turn generated a profound anthropological vertigo, with all the usual symptoms—nausea, disorientation, dizziness, anomie, anxiety—now inscribed on the global human horizon. In our so-called computer age, we have at any given moment dozens of scenarios for the future at our fingertips, all too many of them totally unfit for human life in any recognizable form. There is, admittedly, something arbitrary about choosing to concentrate on one such futuristic catalogue of horrors rather than on another equally plausible series, but certain themes have become increasingly prominent, and should at least be noted:

- It should be known what damage even a single nuclear weapon can do. The best evidence comes of course from Hiroshima and Nagasaki, but this information is by now severely dated.[122] We currently have no weapons in our strategic arsenals as small as the bombs used in 1945.[123] There have, however, been several quite credible recent attempts to marshall the facts we do have in a responsible way. Ground Zero and the International Physicians for the Prevention of Nuclear War[124] have concentrated attention on the damage to be expected in major cities from the explosion of a one-megaton bomb—a minimum figure, since major cities are customarily targeted with 50 to 100 megatons. This literature, along with government damage estimates,[125] is readily available and much of it has been tailored specifically to classroom use.[126]

- It should be known that every city in the U.S. and the U.S.S.R. with a population of 25,000 or more is now—today, this minute—targeted with one or more nuclear weapons.[127]

- It should be known that a single submarine carrying Polaris missiles (not the new beefed-up Tridents) has sufficient firepower to destroy every city in the Soviet Union of 100,000 or more.[128]

- It should be known that for more than 20 years there have been many more missiles than cities on which to target them, so that new generations of missiles can only be targeted at other missiles. The MX missile, for example, can only be construed as a first-strike weapon meant to hit Russian missiles before they are fired, so that all the talk about "safe-basing" and the "hardening" of missile silos is a fairly transparent smoke screen (like its "newspeak" label, *Peacekeeper*) for its only function.

- It should be known that the overkill represented by the current world total of roughly 50,000 nuclear weapons can be expressed in human terms: it is equivalent to about five tons of TNT for every man, woman, and child on the planet.

- It should be known that an all-out nuclear holocaust might well take less than an hour to render at least the northern hemisphere uninhabitable for upwards of a thousand years.

- It should be known that in the late 1980s, the most probable scenario is that of *accidental* nuclear holocaust— either by human error, computer malfunction, or the idiocy of installing automatic satellite "launch-on-warning" systems which would permit no human input at all on the final decision whether our species is to live or to die.[129]

- It should be known that orbiting laser weapons will *never* provide a foolproof defense against all nuclear

weapons, nor is it very likely that such a complicated (and
prohibitively expensive) weapons system could itself ever
be adequately defended.[130] Further, and contrary to the
sanitized image of a "non-nuclear" shield that President
Reagan has tried to promote, the most workable Star
Wars system funded so far would indeed require the
deployment and detonation of nuclear weapons in
space. And who's going to build this marvel? The same
people at Lawrence Livermore Laboratories and Los
Alamos who gave us the hydrogen bomb, and the neutron
bomb, and every other item in our current nuclear
arsenal. Moreover, in the heady atmosphere of its
current efforts to monopolize NASA for the sole purpose
of more speedily militarizing the heavens, the Pentagon
acts as if it expects people to forget that no weapon can
ever be guaranteed to remain *only* defensive. Once you
have weapons in space, of whatever sort, you can aim
them at practically anything and deliver them to their
targets in record time. The Russians are of course
capable of perceiving the obvious, and thus the Reagan
administration's Strategic Defense Initiative has overnight become the single biggest stumbling block to
international arms control. This Star Wars fantasy must,
however, be counted a formidable public relations coup.
Americans tend to be rather easily mesmerized by the
idea of quick technical fixes for complex human
problems. And, at a stroke, the administration has very
nearly succeeded in propelling the nuclear out of the
democratic public forum and back into the bailiwicks
of its highly paid technical experts, who in turn hasten
to assure us that they are the only ones entitled to
an informed opinion on weapons systems that do not
yet exist.

• It should be known that the electromagnetic pulse (EMP) from a single H-bomb detonated in the stratosphere above, say, Omaha would burn out most of the computer and radio and telephone circuits (all fragile silicon filaments these days) across the entire continental United States. It is therefore likely that the missiles which depend on this delicate microcircuitry will, if ever fired, be buzzing around hither and yon completely out of anybody's control.

• It should be known that there are really a great many fingers on "the button." Launch authority for U.S. missiles supposedly belongs to the president, but what if he's off horseback-riding or killed in a surprise attack? Launch authority began to be delegated—to the NORAD commander, the Pacific commander, and the NATO commander—as early as the sixties. It has now filtered down to the point where any nuclear submarine commander, if he loses radio contact with his superiors during what is considered a crisis, has the authority to fire his nuclear missiles at will.[131]

• Of radiation, it should be known that from the mining of uranium (which produces radon gas and exposes small mountains of low-grade uranium ore and tailings), to the processing of weapons-grade uranium, plutonium and tritium, to the testing and incessant retooling of nuclear weapons, to the disposal of radioactive waste (an enormous problem with no solution in sight even if we never build another bomb), every phase of the nuclear weapons production cycle is hazardous to human health in ways and degrees which do not usually show up for over 20 years.[132] The dangers posed by ubiquitous radiation are among the common threads

linking opposition to both nuclear weapons and nuclear
energy production; after all, you cannot make an atomic
bomb without first building a nuclear reactor to produce
the materials. Hence the "Atoms for Peace" propaganda
of the fifties, which provided a thin patina of wholesome-
ness for that decade's crash program of governmentally
subsidized reactor construction. How many more
disasters like Three Mile Island or Kishhtym or
Chernobyl is it going to take to fully alert people to the
virulence of this invisible killer, and to the permanent
threat it represents to future generations? There are
experts in the field who steadfastly maintain that,
beyond the faint traces of background radiation always
present on our planet, there is *no safe level* of exposure to
radioactive emissions. Dr. John Goffman, who first
detected the strontium-90 and iodine-131 today present
in every American mother's milk, has declared any
public exposure to atomic fission products to be "random
premeditated murder."[133] The statistics tell us somebody
will die, they just don't say who. The randomness
associated with expected percentile increases of cancer
incidence in no way mitigates this factor of premeditation:
it has long been evident that the production of nuclear
weapons has exacted and will continue to exact human
casualties, even if the weapons are never used.

• It should be known how much we do *not* know about
the possible effects of a nuclear holocaust. Government
prognostications tend to concentrate on immediate blast
damage to industrial and military targets, ignoring what
they prefer to call "incalculable" effects on the environment at large or on future generations.[134] The nearest
historical precedent for the kind of poisoned world

a holocaust might leave in its wake is that of the Black Death in 1347, which killed over half the population of Europe.[135] But even the Black Death (which began, by the way, as an act of war: plague-infested corpses were catapulted into a beseiged city[136]) is not really even comparable in extent or severity to the horrors that might be spawned by massive radiation poisoning. If there are any survivors, then Nikita Khruschev summarized their fate more than two decades ago: "The survivors will envy the dead."

• It should be known that among the "incalculable" effects of an all-out, or even halfway, use of the superpowers' nuclear arsenals will be a certain (or uncertain) degree of damage to the ozone layer high in the Earth's atmosphere, which serves to screen out harmful excesses of ultraviolet radiation from the sun's fusion furnace—an atmospheric eggshell without which organic life in the terrestrial biosphere would shortly cease. Jack Geiger has estimated that even a partial use of the world's nuclear stockpiles would dissipate 70 percent of the ozone screen in the northern hemisphere, and 30-40 percent in the southern hemisphere.[137] He points out that a mere 20 percent ozone reduction would permanently blind all the animals on the planet. Unable to see, they would naturally be unable to find food, and they would die in short order. While we humans might reassure ourselves that we could wear sunglasses for the rest of our days, the entire food chain would simply begin to die out from under us—beginning with the phytoplankton near the surface of the oceans, the organisms most vulnerable to ultraviolet radiation.

• It should be known that in the past couple of years, Carl Sagan, Paul Ehrlich, and a high-powered research group have popularized the image of a "nuclear winter" following the explosion of any more than about a thousand nuclear weapons, a meteorological catastrophe which might well render academic many of the longer-term "survival" scenarios.[138] Regrettably, planet Earth has only one life to give for our country.

• It should be known that "limited nuclear war"—limited to Europe, the current U.S. Administration would have us believe—is plausible only to those who consider nuclear weapons to be more or less equivalent to conventional weapons. This is a misperception. Once the conflagration starts, it is virtually certain that some of the sparks will get back to the powderkeg and set off the massive arsenals of the U.S. and the U.S.S.R. Once a nuclear weapon is used again in our volatile world, it is a safe bet that it won't take another 40 years until the next use, and the next, and the next. Meanwhile, limited nuclear war scenarios have proven remarkably effective as a rationale for new weapons procurement.

• It should be known that even if the spark never does get back to the powderkeg, that is, if the occasional use of a nuclear weapon—to close a mountain pass in Iran or Lebanon, for example—is somehow countenanced by the superpowers without retaliation, then humankind will have entered into a new and nightmarish era of barbarism and nuclear terrorism where the mass murder of 100,000 or even 500,000 people in a single day becomes business as usual for the "civilized" world.[139] In some ways, all-out holocaust would be preferable to such a new Dark Age. Have we truly reckoned with the

kind of wanton terrorists and monsters we are capable of
becoming?

• It should be known that the danger of the proliferation of
these weapons to nations other than the six present nuclear
powers (namely, the U.S., the U.S.S.R., England, France,
China and India) has never been greater. The commercial
marketing of nuclear reactors all over the world has
made the materials increasingly accessible. Israel and
South Africa apparently have a few (unassembled?)
nuclear devices of their own already. Pakistan claims to
have produced its first bomb early in 1984. Libya is doing
its damnedest to bankroll production of an "Islamic
Bomb." And these are nations which surely have far
fewer compunctions than the superpowers about using
such weapons if pressed into a corner.[140]

• It should be known, as R. Panikkar has stressed, that
the "Third World War" is already upon us, and that the
feared thermonuclear World War III will be only the
inevitable conclusion to the open combat presently
going on almost everywhere in the so-called Third
World.[141] Since World War II, there have in fact been
about 160 wars fought on this small planet, and nearly
40 shooting wars are going on right now. Individually,
these conflicts may seem small in magnitude; collectively,
the death toll already surpasses the estimated 40 million
who died in World War II. In an interdependent world,
it is not a very remote possibility that the direct engagement of the superpowers may finally come as a response
to one of these brushfire conflicts.

• Indeed, it should be well known that as arms merchants
to the world, the superpowers actually have an economic

stake in fomenting and supplying (not to say instigating) the majority of these conflicts. To put a sharp point on it, if the United States stopped selling arms, its entire economy would collapse overnight.[142] As it is, the final economic collapse may take a little longer, but when there are no more markets (because the rest of the client-world is living at too low a standard to support a U.S. economy based on unbridled growth, for example) then the system will indeed come tumbling down. People seem oblivious to the fact that the world economy today is quite bankrupt, despite the capitalist U.S. and the communist U.S.S.R. each offering to the world their own economic panaceas. It is only a question of how long bankers can finagle and renegotiate those middle-term loans that can never be paid back by "developing" countries before the sham is exposed. In the long run, it may well be loss of confidence in banks, international finance, and money itself that will bring down the present paneconomic regime, and this confidence is fast ebbing in the First and Second Worlds as well as in the so-called Third.

• It should be known, for the sake of comparison, that about 20 million people die of starvation every year, mostly women and children; that's one death every other second.[143] Estimates run about $300-400 per person per year for such people to begin to grow or obtain food regularly, a mere dollar a day a person. By contrast, in their combined defense budgets (nuclear and conventional), the U.S. and the U.S.S.R. spend well over $1 billion every day. Where is our sense of proportion, let alone of humanity? Of course, money is no measure of well-being, and by itself no solution to world hunger. But this planet is capable of feeding all its people—Malthus and Garrett

Hardin notwithstanding—and a little money invested in living instead of killing might help keep world hunger at bay, and as a further consequence, reduce the number and intensity of regional battles for locally scarce resources.[144]

Yet we have other priorities, or at least so it would seem to an observer from some other celestial sphere who was able to view the human family as a whole, a "family" with a real talent and apparent passion for self-destruction. Naturally, insights from fields as diverse as agriculture and political science will play a part in any serious discussion of these interlocked issues. The United Nations has made notable strides over the years, as have various religious groups and the Red Cross. And we have lately seen remarkable spontaneous outpourings, like the recent BANDAID and LIVEAID concerts, in response to specific catastrophes. Yet the hard fact remains that more people die every day due to the indifference of their fellow human beings than from any other single cause.

If we appraise closely the dismal skein of dead-end futures conjured up by the nuclear age, we see immediately that for most people the this-worldly goals of capitalism and communism have by and large supplanted (or co-opted) the other-worldly futures promised by the Judaeo-Christian-Islamic religious tradition. It seems as if, collectively and individually, we are no longer capable of sustaining the optimistic vision of an eternal future or a future eternity. Beneath the loud claims of any doctrinal, political, or ideological value system one can always discern the outlines of an underlying anthropology, that is, a conception of the meaning and fulfill-ment of human life. Both communism and capitalism offer anthropologies—that the human being achieves fulfillment in and through the collectivity, the state, on the one hand, or through an autonomous individuality, everyone for oneself, on the other. As we noted earlier, it is important to recognize that these two interpretations of the *humanum*, or human life in its fullness, not only help to define one another but also tend to mirror one another in grotesque caricatures. Each view of the human being—as a function of the collectivity, or as an autonomous "free agent"—is vulnerable to what it leaves out. The collective vision of human destiny is vulnerable precisely to a single individual who claims absolute power, total autonomy: a Stalin, for example. The individualistic vision of human success is vulnerable to a kind of herd instinct or mob rule that too often tends to turn all these every-one-for-oneself types into a manipulable mass of like-thinking (or unthinking) robots, lemmings who will run the rat race to its inevitable conclusion. Between these extremes, both systems resemble each other in many more mundane respects. Both are materialistic, imperialistic, militaristic, and technologically aggressive. Both systems adhere to economics as the primary gradient of human well-being, and yet neither economic scheme really seems capable of doing what it claims to be able to do, namely, to distribute goods and services equitably and keep the peace.

For our purposes here, it is enough to observe that the rudimentary anthropologies underlying both capitalism and communism are often held to be the only options for living a fully human life. Each claims to offer the system whereby the world will prosper, if only they could eradicate the competing system. Since nuclear weapons represent each system's defense mechanism against the other, we are impelled to ask the obvious question: is there no other way to live a fully human life? Are we human beings stuck in such an absurd dialectical impasse that we have no choice but that between Adam Smith's and Karl Marx's vision of the perfect society?

Enlightened self-interest has turned out to be little more than a license for greed, and the dictatorship of the proletariat looks like it intends to postpone indefinitely the true communist state to which Marx supposed it would lead. Is there no other way to envision the *humanum*?

We have already touched on at least one area where the creative imagination clearly still has free play: literature and the arts in general. Here a myriad of alternative patterns for human(e) relatedness to the three worlds, to ourselves and one another, are fleshed out and made real in our experience of them. It is perhaps this ability of ours to imagine alternate futures, alternative destinies, that offers the last best hope for extricating ourselves from the grim embrace of the *danse macabre* in which the two competing dominant anthropologies are presently locked. At the end of his life, Martin Heidegger put it that the true task facing us today is "to safeguard the mystery of Being and watch over the inviolability of the possible."[145] Giles Gunn, who has lately helped work out much of the common ground between religion and literature, expresses the "hypothetical" character of literature in this way:

> Every work of literature, as a hypothetical creation, presupposes for its very existence. . .a "commitment to vital possibility.". . .Our response to literature, while unavoidably dependent on "our deepest sense of ourselves," is not absolutely determined by that sense. Indeed, it is precisely that sense which literature, in suggesting "that something else may be the case," seeks to extend, to complicate, and ultimately to transform. Which is only to say that at bottom literature seeks to convert by extending the range of our imaginative capacity and hence "our knowledge and governance of human possibility."[146]

Noteworthy also is Gunn's suggestion that it is mainly by the encounter with and assimilation of "otherness" that this "sense of ourselves" is most profoundly transformed.[147] George Steiner carries this intuition beyond formal literature into a reflection on the human capacity for language altogether. In the Afterword to his *After Babel*, Steiner writes:

> I have put forward the hypothesis that the proliferation of mutually incomprehensible tongues stems from an absolutely fundamental impulse in language itself. I believe that the communication of information, of ostensive and verifiable "facts," constitutes only one part, and perhaps a secondary part, of human discourse. The potentials of fiction, of counter-factuality, of undecidable futurity profoundly characterize both the origins and nature of speech. They differentiate it ontologically from the many signal systems available to the animal world. They determine the unique, often ambiguous tenor of human consciousness and make the relations of that consciousness to "reality" creative. Through language, so much of which is focused inward to our private selves, we reject the empirical inevitability of the world. Through language, we construct what I have called "alternities of being."[148]

To summarize, then: over against the grotesquely fore-shortened "futures" with which we are faced when we begin to examine the nuts and bolts of nuclear proliferation, there yet remains the sacred precinct of language, literature, and the fine arts, where human potentialities still retain their primordial vitality. Earlier on, I tried to draw attention to the classical religious texts now accessible—perhaps too superficially accessible—in modern languages. Here one should equally emphasize at least three areas of contemporary literature which may help provide leverage on the "empirical inevitability" of the current human predicament.

In the first place, there is the well-known 20th century literature of alienation, which puts the "no exit" of modern western civilization at the personal level where it is, before and after all, most real. If the nuclear issue is truly human(e) relativity in all its aspects, then we should expect to find its symptomology precisely where human beings relate, or do not relate, to one another most intimately and most passionately. If the "enemy" is truly within us, then works like Melville's *Moby Dick* or Conrad's *Heart of Darkness* (newly incarnated in Coppola's film, *Apocalypse Now*), might have a great deal to tell us about why we human beings so often seem to be bent on destroying ourselves and one another for grandiose reasons and high ideals. On a more positive footing, we must not let the dust collect on the works of the great modern poets—one thinks of Yeats, Pound, Eliot, Auden, Williams, Stevens, Bunting, Zukofsky, Olson, and Oppen—who so clearly celebrated life-enhancing alternatives to cultural suicide. This larger human horizon is currently in danger of being altogether lost, as John Haines has lately lamented, in the inconsequential chatter and self-indulgent gush that passes for post-modernist poetry. This shortfall of language also belongs to the symptomology of the nuclear age:

> The world of the poet has shrunk many times since the days when Wallace Stevens and William Carlos Williams took for their concern the whole of life, or at least the whole life of a place. The world with which the contemporary poet characteristically concerns himself or herself resembles the self-limited world of the adolescent. It is a deliberate limitation that comes, I believe, from despair, as if the meaning of our situation,

the weight of the disasters that threaten us, is so huge
that we cannot find words for it, nor perhaps even
emotions. Therefore we shrink, become deliberately
small and trivial, and chatter about nothing at all,
huddled like apes before a storm.[149]

Secondly, during the past couple of decades an enormous
world literature has grown up as the so-called Third World
peoples have begun to come to terms with the clash between
their own traditional values and the paneconomic technological
System today doing business in the name of western civilization. Many of these works have originally been composed in
English—Wole Soyinka's plays and Chinua Achebe's novels
from Nigeria, for example—a language fast becoming, for
good *and* for ill, a lingua franca for much of the world. It is
here, in the way these "others" (the term is not uniformly
appropriate) view western values and civilization, that we
may begin to learn where our own blind spots are, and where
vital alternatives may lie. It suffices merely to mention the
names of writers like Jorge Luis Borges or Gabriel Garcia
Marquez to suggest the powerful range of the fully engaged
human imagination. Borges' story "The End of the Duel'"[150]
offers, I would suggest, a grisly image of today's nuclear
predicament as the ultimate duel of honor: two condemned
rivals running a race *after* their throats have been slit. The one
who lurches farthest before expiring is supposed to be the
winner.

Finally, though perhaps more for its psychosocial
critiques and insights than for consistent literary merit,
I would recommend more attention to the burgeoning
literature lumped together (and often studiously neglected)
under the "minor" genres of science fiction and certain
sophisticated horror stories, which have from the time of

Mary Shelley's *Frankenstein* registered and probed the deleterious effects of single-minded science. This literature, which often finds its way into popular movies, ranges from the purely imaginary worlds of the fantasists and surrealists to quite pragmatic extrapolations of current trends in physics, biology, and medicine. These kinds of stories have a way of unearthing forces at play in the collective unconscious. In the traditional sci-fi thriller, there will be an eruption of demonic forces from the nether regions—Japanese dinosaurs, alien invaders from outer space, giant irradiated insects, mutant microbes and so forth—and then, following the conventions of the genre, the military-industrial-scientific complex will collaborate to push whatever it is back into the shadow realm from which it emerged. This paradigm held true almost universally in the fifties and early sixties. But in the last 20 years or so of such films, the "conscious" forces of government and science have been notably less successful at dominating these demons from the "unconscious"—which we may take to be an accurate reflection of the true psycho-social state of human affairs. Beyond offering what might be called "possible futures," it should be noted that the most prominent theme of science-fiction stories since 1945 has been atomic catastrophe or misadventure by radiation contamination. Recently, and I think significantly, themes of demonic possession and Antichrist have been taking over the genre. All in all, we find that a full range of responses to the nuclear predicament—everything from the end of the world to the fail-safe fallacy to escapism and survivalist fantasies— have been played out in imaginative ways which are for the most part more realistic than the neat scenarios and antiseptic body-counts produced by the Pentagon and the RAND Corporation's computer strategists. It cannot be denied that what is most often at stake in popular science fiction fare from

Star Trek to *Star Wars* to *Dr. Who* is the urgent question of how to save the world. The question is apparently more real to the post-Hiroshima generation than to their parents, who often find their children's fascination with such stories quite inexplicable.

All of the above are simply suggestions for provoking, on a number of fronts, a real experience of the life-or-death options of the nuclear age in the college and university setting. Whether one approaches through a scrupulous appraisal of the facts of nuclear proliferation, through a fictional exploration of possible futures, or both together, a further proviso is in order: the future can only be what we make of it. "Trend is not destiny," as René Dubos liked to declare,[151] and all possible futures are in our hands today. There is, as R. Panikkar suggests, a genuine anthropological basis for hope in the future:

> The human being is that being which creates its own future. The human future does not exist; the models we entertain of it belong at most to epistemology, never to ontology. Human life on Earth . . . is a true creation, an expansion that has no other law but the freedom of Man, who as he gradually advances creates his own future, his own situation, his being. Freedom is human creativity. Man is free insofar as he creates or, better said, to the extent that he creates himself.[152]

The Present
(Religion and Culture)

If the kingdom of God is always-already-but-not-yet-present,[153] the New Age is always now or never. There is a difference. That is to say, the enhancement of life in the post-Hiroshima world has a certain obvious urgency to it. Today the kingdom of peace will open its gates either to everybody or to nobody. Inaction is culpable, since it perpetuates a predicament in which the world proceeds apace to its own demise. Even if one accepts the traditional myth of an other-worldly paradise, sitting idly by while God's creation goes to the devil, so to speak, could hardly be construed to confer a ticket to Heaven. As noted earlier, religion should not be confused with gnosticism. It is not abstract knowledge that saves, but faith, and works, and love. In this light, even philosophy (*philo-sophia*) is not just the love of wisdom, but the wisdom of love.[154] By and large, the mainstream religions of humankind share at least one common concern: the rescue of

human beings from the danger of perishing utterly.[155] Today
that danger could not be more real, but there has been
a change in emphasis. It is not only individual salvation that is
at stake, but the survival of humankind and the entire reality
of our experience. Consequently, any effort to surmount the
nuclear threat to the whole of Life already has a religious
dimension built into it. And by the same token, I would also
submit that religious studies as an academic discipline is
admirably suited to serve as a "nucleus" for the study of the
very many facets of human(e) relativity which together
constitute the nuclear issue in the usual sense. Indeed, I can
think of no more religious study than what we have set forth
from the outset as "the exploratory strategy of starting with
the whole,"[156] that is, the concerted effort to come to grips
with the many seemingly disparate elements of this complex
issue. We shall examine below some of the questions posed
by the specifically religious study of war and peace, so it
should suffice here simply to outline some of the ever-present
alternatives to thermonuclear holocaust. They are legion,
once you start looking for them. Such topics ought to be part
and parcel of any attempt to deal integrally with the nuclear
issue in the classroom, particularly within the context of
religious education. Even though any list of "what's
happening" here and now is bound to be provisional and
incomplete, we have to make a start. Several directions of
inquiry are immediately apparent:

> • One should pay close attention to the ongoing contem-
> porary discussion of the issue. Nobody wants a nuclear
> war; the question is how to prevent one. Since about the
> fall of 1981, the media have daily covered anti-nuclear
> protests in Europe and the U.S., debates over deterrence,
> intermediate-range missiles, a no-first-use policy (to
> which the U.S. has never agreed), launch-on-warning,

defense budget items, weapons in space, and so forth. Public broadcasting stations and educational television in particular have made commendable efforts to bring the issue into every American home. There is no substitute for attention to and participation in public discussion of public policy.

• One should pay close attention to arms negotiations of any and all sorts, whether in Geneva or in the United Nations or in any other forum. The U.N. has published extensive documentation of its own discussions over the past 39 years.[157] The endlessly frustrated effort to create an international agency for the control of all nuclear weapons and materials began with American Ambassador Bernard Baruch's proposal to the fledgling United Nations in 1946: "We are here to make a choice between the quick and the dead."

• One should pay close attention to the vicissitudes of the nuclear weapons' freeze campaign and related efforts in the world at large as well as in the more limited arena of American politics.[158] Obviously, a moratorium on production and further deployment is a good start, but there are myriad questions involved here which cannot be answered in simplistic terms. And what happens when the nuclear issue, human(e) relatedness as a whole, is reduced to merely one more political option on the ballot, to which we are merely supposed to say yes or no. Is that it? What next?[159]

• One should pay close attention to the recent growth of peace academies, institutes for conflict resolution, and the like. Almost every major university now has some sort of peace studies group and can usually provide spokespeople to address a class on local activities. The proposed United States Peace Academy may so far amount only to window dressing for current belligerent policies, but the idea is an intriguing one and bears watching.

• One should play close attention to nonviolence and pacifism in all its forms, its (sociological) history and (psychological and theological) motivations. In the final analysis, violence in the family setting, abortion, crime in the streets, and abuse of the environment cannot really be divorced from the nuclear issue: human relations of every sort are at risk here. It should be clear that there are workable alternatives to violence at every level of human society, and that conflict models do not exhaust the repertoire of human potentials for relationship. It is extremely fortuitous, for instance, that a film of the stature of Attenborough's *Gandhi* should emerge at this precarious juncture in human affairs. Whether or not the film's portrayal is an accurate rendering of the life and times of the original Gandhi becomes a secondary issue. This new Gandhi is already having his own direct effect on a generation which never saw the original in the newsreels of the forties and fifties. And new questions are arising about the propriety and effectiveness of civil disobedience in different cultural contexts—whether Gandhi's tactics would work against a Hitler or a Stalin, for example.

• Along the same lines, one should pay close attention to the growing library of audiovisual resources on the nuclear issue now available for use in the classroom. Many organizations are now providing professionally produced tape recordings and video cassettes which can easily be adapted to stimulate discussion on a variety of topics. And as noted, public and even commercial television stations are lately tending to devote more and more air time to such issues.[160]

• One should pay close attention to cultural exchange between the superpowers, especially in this recent period of "thaw." Until the summit in 1985, the U.S. government had managed to cut off very nearly every form of cultural intercourse between this country and the Soviet Union. Now more than ever, we would be well advised to take advantage of every opportunity to get to know our "enemy" and address the differences between us through less official channels. In the year 1983, 12-year-old Samantha Smith was arguably the most effective diplomat in the area of U.S./Soviet relations. After her tragic death in late 1985, American officialdom seems to have forgotten her efforts to prod the superpowers toward a more peaceful world. Not so in the Soviet Union. As *Time* magazine recently reported, "In the U.S.S.R., a diamond, a flower, a street, a poem, and a book have already been named in her honor. Now comes a Samantha Smith stamp, worth 5 kopecks, or about 7.5. . . . Schoolchildren in Tashkent have formed an international friendship club and set up a museum in her memory, and students from more than 100 Soviet schools competed for the chance to have their school renamed Samantha Smith."[161]

• One should also pay close attention to every imaginable kind of alternative lifestyle, from new forms of monasticism to counterculture communes.[162] Handbooks of appropriate technology[163] and guides to various spiritual communities[164] are increasingly available. Many very healthy efforts to expound a new world order are afoot and out in the

open today, ranging from the U.N. Law of the Sea
conferences to networks for "second-track" citizen
diplomacy to psychological and sociological studies of
human and governmental institutions.[165] Donald Keys,
founder of the exemplary Planetary Citizens organization,
writes, "There is no commitment in the U.N. Charter to
develop supra-national institutions. The Charter model
is basically a nineteenth-century 'concert of powers'
model, based on the sovereignty of States. . . .The U.N.
shows clear signs of ultimately outgrowing that model.
In many respects, the U.N. . . . is still a highly transitional
institution, not yet the focus for functional world
community which seems to be required by the global
pressures of today."[166] One theme not often enough
addressed in world order discussions is whether any
human order (institutional, political, economic, or
religious) ought properly to be imposed as a kind of
superstructure on planet Earth, or whether there is in
fact already an existing planetary order to which we
humans should rather learn to orient or attune ourselves
directly.[167]

• Finally, one should also pay close attention to the most
ancient healing arts. In the rhythms of music, in the
proportions of architecture, in the patterns of making
a mandala of whatever sort, we recover the acute attention
and focused energy it takes to find harmony among
disparate elements, and to bring forth a cosmos, an
ordered whole, out of ambient chaos. Most scholarly
studies tend to be analytical in the sense that we break
down our objects of study into their components. Now,
with the doomsday clock always ticking just out of earshot,
we must also emphasize creativity in all its forms, we
must re-learn how to put the world back together again.

The arts have always been the natural/cultural vehicles for most effectively realizing the physical/metaphysical coherence of the whole. As José Argüelles points out, "If art is no longer specialized, then art becomes a means of relating to the whole; that is, it becomes an activity that responds to and helps direct environmental impulses rather than an art (or a technology) that is imposed on the environment."[168]

* * *

All in all, it must be conceded that the issues of war and peace between and among the bickering children of the human family have never been very clearly understood. It is now or never. Human relativity has an ominous side: we humans may be so relative a phenomenon that we can pass into oblivion in the blink of an eye, leaving nary a ripple in the maelstrom of cosmic evolution. Even should we have to jettison our conventional frames of reference in order to pursue the study of alternatives, we face here the first and most obvious imperative on the human agenda today: we must learn to relate—to ourselves, to one another, and to Heaven and Earth as well—creatively and constructively, for a change, and damned quickly, before we destroy ourselves either by design or misunderstanding or computer error. This means closer and deeper and more scrupulous probes into at least three areas: peace studies, war studies, and the institutionalization of human values. Let us look briefly into these areas one by one.

Peace Studies

Would we really know peace even if we stumbled into it? How many Americans know the Russian word for peace?[169] Is it not indeed high time to undertake peace studies in earnest at the university level?—if only in order to dispel the all-too-common notion that peace is some sort of fluke in human affairs, a lull between the wars that chequer our history, a vague and amorphous state of mind or of culture which arises only when exhaustion debilitates the "normal" strife between people and peoples. When we look more closely, three or four distinct sorts of peace call for our interrogation.

Military peace is of two types:

• The truce, the forced calm, the pause between battles; it usually represents the attempt to regroup for further fighting, or to negotiate a better fighting position. The cold war and the MAD policy of deterrence are examples of this kind of anxious instability prolonged to the point of cultural psychosis.

• The victory, triumph, conquest, is the ideal form of military peace; it is an enforced tranquillity. And yet even victory usually proves unstable, since it is generally only a prelude to further conflict once the vanquished regain their strength and begin to strike back.

Neither the uneasy calm of the truce nor the volatile coercion of victory can really qualify as peace for the nuclear age, though these are the categories most commonly mistaken for a true peace. There is assumed to be a brutal realism underlying these military concepts of peace, but in fact the generals are always thinking in terms of the last war, and the validity of such military realism has been all but annulled by the current global predicament. The truce of mutual nuclear terrorism (deterrence) is untenable and explosive both politically and psychologically, and victory inconceivable in any actual exchange of nuclear weapons. What we have is not peace, but conflict postponed. . .until?

Political peace corresponds to the etymology of the word in the Latin-based languages. Here peace is viewed as stemming from *pacem, pactum*: a pact, a compact, an agreement. The usual form of such a pact is an agreement between the conflicting parties to stay out of one another's way or vital interests: "You leave us alone, and we'll leave you alone." It is a static peace, or rather, an attempt to end conflict by engineering a political stasis. But neither human nature nor human culture are static, and political peace is notoriously susceptible to erosion by shifting values, geopolitical contexts, and ideological priorities of the parties involved. There is a naive idealism underlying most attempts at merely political peace. In this regard, it should be noted that there are presently on record about 8,000 treaties of peace in western history, most of them including a phrase avowing that a lasting—nay, eternal—peace will ensue upon the signing of that particular document.[170] Political peace is, however, a rather superficial stasis; it resolves the symptoms of overt conflict, but too often leaves the causes untouched. It is peace for the time being, and can (like the Versailles treaty) as often as not merely set the stage for further conflict. The answer must lie deeper. It should be noted that the founders of the United Nations were well aware of the limitations of the political peace described here and indeed took pains to underscore in the Constitution of UNESCO (1945):

That a peace based exclusively upon the political and economic arrangements of governments would not be a peace which could secure the unanimous, lasting and sincere support of the peoples of the world, and that the peace must therefore be founded, if it is not to fail, upon the intellectual and moral solidarity of mankind.

For these reasons, the States Parties to this Constitution, believing in full and equal opportunities for education for all, in the unrestricted pursuit of objective truth, and in the free exchange of ideas and knowledge, are agreed and determined to develop and to increase the means of communication between their peoples and to employ these means for the purposes of mutual understanding and a truer and more perfect knowledge of each others' lives.

Religious peace is of another kind altogether. It is based on the mutual recognition that there are nonmanipulable factors in human life which neither party can monopolize. It is undergirded neither by realism nor by idealism, but by the power of one or more living symbols. Today peace has become just such a living religious symbol, uniting many formerly disparate movements and concerns. Properly understood, religious peace has both objective and subjective facets, and cannot be reduced to either one without the other.

  • *Objectively,* religious peace comes to light in the constructive dialogical encounter of people who hold radically dissimilar worldviews.[171] It is not a stasis, but a dynamic integrity of tensions. Dialectical conflict is transformed into a creative polarity, crackling with energy. It involves a positive acceptance of human diversity, as specified earlier, and a tolerance of what may from an ideological viewpoint seem to be intolerably incompatible worldviews. The present-day encounter of formerly insular religious and cultural traditions offers the possibility of cross-fertilization, but it is rooted in common sense. In the thermonuclear arena, you cannot destroy "the others" without destroying yourself. Therefore, though it may often be a "hot" peace and a struggle, we are going to have to learn to cope with the existence and sometimes "alien" ideas of our fellow human beings. Intercultural religious studies are thus a sign of hope in a world where international relations are daily deteriorating.

• *Subjectively*, we must recognize the real value of a contemplative attitude—a mystical peace, if the word is understood aright—which implies much more than some bland state of alpha brainwave transmissions. The spiritualities of the major religious traditions offer us plenty of precedents, models, and exemplary figures, but the context has changed radically. The shape spiritual disciplines will take in the nuclear age is a crucial subtopic which will be addressed more closely in the "Metalogue." Why should spiritual disciplines matter to peacemakers? Because the contemplative stands in the "free space," the other side of space and time, where opposites coincide, where paradox reigns, where the truth itself is tolerant of human errors and shortcomings, and the mystery of Life intelligible—precisely as mystery. And today, as R. Panikkar has lately stressed, "It is no longer some few individuals who may reach salvation, i.e. true happiness, the fulfillment of life or whatever, by crossing to the other shore, experiencing the transtemporal, the tempiternal. It is Humankind itself which is now making this breakthrough in its consciousness out of sheer survival necessity—because of the stifling closeness of the System. It is precisely the instinct for survival that today throws us toward the other shore of time and space."[172]

From an integral perspective, it should be noted, there is really no difference whatsoever between the subjective and objective facets of the religious peace-in-the-making sketched above. Peace comes to pass in and through the tabernacle of the person, who is "nothing but" the totality of relations with Heaven, Earth, and every other person. We might add, drawing again from Panikkar, that an intra-religious dialogue accompanies and reinforces all inter-religious dialogue.[173]

* * *

Obviously, any such sketches must remain extremely tentative. These are topics which cry out for further study, questions which require our full and unflinching attention. The long and the short of it is that peace will not be achieved by luck, or accident, or happenstance. One can do little more in such short compass than to suggest possible pathways for further exploration and extrapolation. The following might most plainly be called the "common sense" approach to dwelling peaceably with one another.

"Common sense" in its most common sense means using your head, and the medieval understanding of the mind as the *common* sense (integrating input from the five physical senses) stands behind this notion. Indubitably, it applies. There is however a deeper common sense, that of all the words for human conviviality that are etymologically rooted in *common*. The word *common* has two Latin roots: *cum*, which means "together," and *munis*, meaning "ready to serve" (derived from *munus*, to serve). Thus the common denominator of the words (and thereby the accumulated human experience) we know as *community, communication*, and *communion* is this shared sense of being "ready to serve together." (To this the philologist Skeat adds, "as if 'helping each other'."[174]) So the common sense here is also the sense of *the commons*,[175] the perception of the true common-wealth of humankind.

Now "humanity" too has a double meaning. It can mean "the human beings," as distinguished from rocks, plants, animals, angels, and Gods. But it also, and perhaps more fundamentally, means being humane. Thus humanity denotes an act which our common sense ought to tell us is at once the goal and the basis of any attempt to be authentically human: caring for, instead of trying to kill, one another. Let us try to unravel this primordial human self-understanding under the threefold rubric of community (as the *locus*), communication (as the *modus*), and communion (as the *nucleus*). All together make up what we are calling human(e) relativity.

Community

Deference, not dominance, constitutes human community. Deference does not mean acquiescence or submission, but rather reciprocity and mutual respect. Think of the hallowed courtesy of the "polite" societies of China and Japan. The ill-mannered nation-state, by contrast, is founded mainly on force. What cripples our sense of shared community in the modern world is most likely the tendency to place all our social emphasis on active virtues—we want to grasp, to know, to discover, to love—while in fact human community depends in great part on more passive virtues. We forget that it is at least equally important to let oneself *be* grasped, known, discovered, and loved by others. True community is founded in common, on the readiness to be there for one another. Here quite homely values take precedence over trumped-up ideologies. Convivial tools and activities contribute to the subsistence of the commonwealth, as Ivan Illich ever insists,[176] while industrial, bureaucratic, and military super-structures merely clutter up its workings. Even the city as centralized megalopolis has proven to us its economic, cultural, and ecological unviability, and the nation-state certainly no less so, as the works of Lewis Mumford cited earlier make all too evident. Most of us do not really form our communities solely on the basis of physical proximity; we tend rather to form "communities of interest" based on friendship, collegiality, and mutual service. On a global scale what seems to be needed is a network, or many interlocking networks; a loose-knit planetary community of communities, and not some new world empire or super state.

Here then is the locus of the common sense approach to dwelling in a world bristling with 50,000 nuclear weapons. We are all under the gun: the inversion of present doomsday trends begins at home, in the most basic realm of neighborliness, friendliness, and conviviality.

Communication

To begin weaving a durable and decentralized intercultural fabric, we need above all to keep in touch with one another. The key here, as we have stressed, is dialogue, not dialectics. We are all too accustomed to one-way modes of communication: the loudspeaker, the computer printout, the television. We tie ourselves into knots trying to squeeze all human intercourse into the yes/no, either/or, profit/loss, win/lose strictures of dialectical conflict. It seems we like to argue. Dialogue, on the other hand, assumes only that nobody has a monopoly on the ultimate human values, and that everybody has something to learn from everybody else. Here the dialect of the tribe, the regional vernacular, and the untranslatable idiom are often the most appropriate vehicles, while legalese, bureaucratese, technical jargon, and political cant merely obfuscate the human issues in any genuine effort to understand and be understood. In our era of data overload, we should long since have realized that the quantity of standardized and preprocessed pabulum transmitted from one mechanical terminal to another has very little to do with the quality of true human intimacy. Is there not room aplenty in our many interpenetrating universes of discourse for anyone and everyone to say their own word, dance to their own tune, or draw out their own vision, without renouncing mutual intelligibility?

This then addresses the mode of articulating a global transvaluation of values, from bilateral deception to mutual candor: speak unto others as you would be spoken to, and try to listen likewise.

Communion

At the heart of human(e) relativity there is the un-understandable mystery that allows human beings, despite all odds, to understand one another at all. It is nothing short of astonishing, with about four thousand mutually incomprehensible tongues currently wagging around the globe, that we are able to translate, assimilate, and understand as much as we do. If one had only all the babble and the dictionaries to go by, these diverse linguistic and cultural worlds crisscrossing our planet would seem calculated solely to maximize misunderstanding. Fortunately, there's more to it. There is—because there must be—a common ground, an open nexus of human relativity, a clear center which permits us to see through our (cultural, linguistic, religious, and ideological) conditions and conditionings in order to experience the reality of the other. Here we can only underscore this dynamism of the living dialogue, which is in essence utterly free. If there were not always already this communion, if we did not already "stand under" that primordial mystery which permits us to understand one another, then we would stand little chance at all of surviving our own instinctual and institutional defense mechanisms.

Here then is the nucleus of human(e) relations, the common origin of understanding, which none of us may possess or manipulate with impunity. Together we stand already at this place of peace, which is no "place" at all, but rather *the awareness* that there is between us something which surpasses us: that mystery we call Life.

If we do not cultivate this common ground, our very humanity withers. If we fight over this our commonwealth, we shall surely vanquish only ourselves. If we do not act out of our common sense of the shared treasure of Life, we become not human(e) beings but merely wanton destroyers, worthy of whatever dismal fate we would mete out to others.

But hope is faith in love.[177]

War Studies

We should say it in so many words: peace will not be achieved by means of war. The development of nuclear, chemical, and biological weapons of biocide has rendered the old rationalization of "the war to end all wars" (logically shaky at best) not only absurd but lethal. If, on the other hand, our peace studies are not to remain only unrealistic dreams we can no longer shy away from grappling with whatever it is that repeatedly impels human beings to undertake the systematic slaughter of their own kind. War studies are a necessary ingredient of peace studies, and a kind of shadowy underworld seething beneath all the sunnier horizons of religious studies. The cultural anthropologists are far ahead of most students of religion in coming to grips with the ugly side of human relations—war, violence, fanaticism, aggression, persecution, etc.—although these are often directly and thematically religious manifestations.[178] The reluctance is understandable, of course, but it is time to stop pretending that religion has no complicity in these collective aberrations.

Among the most intractable questions which must be addressed with all the resources and perspectives available to religious studies are the following:

Why War?

This question, which formed the title of Freud's famous letter to Einstein,[179] has never really been adequately answered. The mainline academic disciplines which regularly attend to the issue—that is, history, political and military science, strategic studies, sociology, psychology, and cultural anthropology—all hold pieces of the jigsaw puzzle, but the whole picture of organized warfare as a concerted human activity has yet to be assembled. There are, however, several clues which ought not to be ignored in any religious inquiry into nature and origins of war. We know, for example, several things that war is not:

War is not instinctual. In attenuated form, humans have the normal mammalian reaction of "defensive aggression," the so-called fight-or-flight reaction that pumps adrenalin into us when we feel threatened. Now there is a school of thought, first systematically set forth in Hobbes's *Leviathan* and today perhaps best represented by G. Hardin, which holds that the sovereignty of the nation-state is the only "rational" bulwark against "natural" human aggressiveness[180]— in short, a sociological version of the "killer ape" theory.[181] But the best evidence today seems to indicate, in sharp contrast, that we are culturally—and not physically or naturally— predators. We have evolved our most lethal instruments rather recently, and so have not had sufficient genetic evolution behind us to acquire the customary instinctual safeguards which inhibit true predators from killing their con-specifics. Wolves, for example, will fight only until the weaker concedes defeat by exposing its jugular to the stronger: end of contest. But two non-predators—caged doves, for instance, fighting over limited food—will indeed fight to the death.[182] No, the fight-or-flight reaction is not in itself sufficient to account for warmaking. The real problem is that this otherwise adaptive mechanism responds in the same way to both genuine threats and illusory ones. Here, as with any activity specifically human, the centrality of language and symbol comes to the fore: propaganda, slogans, ideologies of all sorts, as well as outright lies can convince an individual or a community it is being threatened. As observed earlier, Jung's studies of the "shadow" and of the lethal paranoia generated when we project our own propensities for evil or destruction onto some halfway hallucinated enemy should long since have illuminated the way our basically healthy instinct for self-preservation can be rechannelled into wanton rampages of destruction. The dark and delinquent side of the human psyche has the potential for doing incredible damage while our conscious opinion of ourselves remains undisturbed in its complacency. The answer that almost any soldier in the midst of any war would give to Freud's question is appalling in its simplicity and terrifying in its irrationality: "Why are 'we' fighting this war? Because 'they' forced us into it; it is 'their' fault." The psychopathology of violence shows itself plainly in the many ways we try to dodge responsibility for our own destructiveness. And if sanity seeks to effect changes at these depths of the psyche, Jung's warning at the height of the cold war mentality of the mid-fifties still stands: "Reason alone does not suffice."[183]

War is not a universal human phenomenon. Contrary to many 19th century European preconceptions about the savagery of people assumed to be primitive, neolithic village cultures generally—and most of the contemporary tribal cultures investigated over the past hundred years—disclose no such massive cultural preoccupation with organized murder as do the supposedly civilized nation states. Indeed, organized, long-term warfare may well be a psychosocial disorder or disease endemic to civilization in certain stages. More specifically, the rise of warfare seems to be part and parcel of the centralized city, state, nation, and empire-building of historical peoples. Erich Fromm puts it laconically: "War as an institution was a new invention, like kingdom or bureaucracy, made about 3,000 B.C."[184] Stanley Diamond is equally direct: "Civilization originates in conquest abroad and repression at home."[185] Both, evidently, agree with Lewis Mumford's characterization of the negative features that accompany the transition from village culture to city-culture:

> Out of the early neolithic complex a different kind of social organization arose: no longer dispersed in small units, but unified in a large one; no longer "democratic," that is, based on neighborly intimacy, customary usage, and consent, but authoritarian, centrally directed, under the control of a dominant minority; no longer confined to a limited territory, but deliberately "going out of bounds" to seize new materials and enslave helpless men, to exercise control, to exact tribute. This new culture was dedicated, not just to the enhancement of life, but to the expansion of collective power. By perfecting new instruments of coercion, the rulers of this society had, by the Third Millennium, B.C., organized industrial and military power on a scale that was never to be surpassed until our own time.[186]

War is not inevitable; it requires certain underlying cultural assumptions. R. Panikkar has sharply underscored one such assumption: "History leads inevitably to war."[187] By history here he presumably means the prevailing *mythos* of the Judaeo-Christian-Islamic-Marxist tradition, and its powerful attempts to resolve global human dilemmas within the structures and strictures of a single civilization organized on the monolithic (and usually monotheistic) imperial model. This raises the

question of whether and to what extent history itself, as the
collective myth of the various "chosen" peoples, inveterate
squabblers all, must inevitably produce war after war until it
all comes to an end—probably with a bang and not with
a whimper. Given today's thermonuclear arsenals, the end of
history is certainly threatening to bring about the end of
humankind and the Death of the Earth along with it, although
history and historical consciousness are epiphenomena in
these larger contexts. The question is still open, on the other
hand, whether we have the recourse to address and transform
the cultural assumptions that produce wars—in short, to
consciously begin to bring history to a close so that humankind
and the Earth may survive its passing. All possibilities converge
on the present moment: now or never.

What's so sacred about sovereignty?

Jonathan Schell strikes the keynote here: Disarmament
implies desovereigntization,[188] and presumably vice versa.
The most basic claim of sovereignty, more fundamental than
coining money or collecting taxes, is the claimed right to
resort to warfare to preserve itself. The symbol of sovereignty
is the sceptre, in fact a glorified mace, an instrument useful
mainly for bashing human heads.[189] In any given war, no
sovereign state is obliged to surrender until it has used all the
weapons at its disposal. And, in contrast to municipal, natural
or divine law, the preservation of the nation-state is presently
considered an *absolute* right under international law. In the
nuclear age, this volatile combination of factors can result only
in holocaust. Today, therefore, the sovereign nation-state
appears to be a malignant atavism ready to sacrifice most of
the people on the planet in order to safeguard its national
security. In a way, the question is not whether the nation-
state will survive—we can be assured that nations will do

everything in their considerable power to preserve themselves—
but whether humankind will survive the survival of the nation-
state. We humans are the mortal ones. Conceding, then, the
obsolescence of state sovereignty in the nuclear age, several
important areas of study have too long been overlooked, and
require our immediate attention:

Human sacrifice. Sovereignty is an institution built on mass
human sacrifice. Now most religions center on the total conversion
or transformation of the human person, and thus call for some
kind of interiorization or sublimation of the dynamism of
sacrifice. It is understood that the true sacrifice is not to die,
but to live—to give one's life in this positive fashion by living
it for others, for God, for truth.[190] So how did this salutary
human sacrifice (of the ego, of the passions, and so forth) get
turned so violently around? The truly religious sacrifice is
accomplished without homicide or suicide, as in the eucharistic
sacrifice of the Christian Mass. How then did entire cultures
get the idea that the atrocious killing of another human being
was going to do them any good whatsoever?

It is speculative, but probably the moment comes when it
is not I who has to change, not I who has to make the sacrifice,
but you. You're to blame, you're responsible. If we get rid of
you (even if I, too, risk destruction in the process), all will be
well again. We recognize here a familiar pattern. First, disorder
in the community; then, the attempt to fix blame; and finally,
the discovery (or rather invention) of a scapegoat who, so to
speak, bears the sins of the community, and is either
slaughtered on the spot or driven out into the wilds to die.
Gradually, the entire procedure is institutionalized. Scape-
goats are sought outside the community, captives are taken
and sacrificed, recriminations follow, more deaths, more
senseless slaughter, and eventually you have an entire society
organized around collective human sacrifice. Of course, such

a sequence can only be sketched in broad strokes from thousands of distinct and separate incidents, but the cumulative effect, unceasing conflict between human groups, is plain enough in our contemporary world.

Some scholars trace this perversion all the way back to the prehistorical agricultural setting, where the blood of a human victim is supposed to insure the fertility of the crops. Others, Joseph Campbell for example, assert that tropical tribes, seeing all around them new life arising from decay in the vegetable world, first drew the unwarranted conclusion that killing increases life, and that to activate life one must kill.[191] Campbell here echoes the findings of Mircea Eliade, who had in the fifties drawn attention to the myths of early horticulturalists in the tropical region in which "the edible plant is not given in nature; it is the product of a primordial sacrifice. In mythical times, a semi-divine being is sacrificed in order that tubers and fruit trees may grow out of his or her body."[192] Similar warrants may underlie the myths of Osiris, Tammuz, and Adonis in the ancient Near East, and were certainly explicit in the wholesale prodigies of human sacrifice practiced by the Maya and the Aztecs in Mesoamerica. In much the same vein, as Lewis Mumford reminds us, "the sacrificial offerings of fruits by Cain, the farmer, were less acceptable to Jehovah than those of Abel, the pastoralist, who sacrificed an animal."[193] Cain's response—the sacrifice of his brother, Abel—may have been out of proportion, but it is a logical response and could in fact be interpreted as simply an excess of zeal to please a God who seemed to favor blood sacrifice.

Still and all, the sacrifice of a human victim to insure fertility is a long way from open warfare. It must be admitted that this special form of mass human sacrifice is nearly always specific to the parasitic institution of *divine kingship* in the historical period. Indeed, the king is often held to be the ultimate scapegoat. If things go badly, the king must die. But kings are notoriously adept at turning tables on the scapegoat mentality, and turning it to their own advantage. As the leader, the king is in a prominent and powerful enough position to point the accusing finger at somebody else and shift the blame, a phenomenon we see in all its blatancy during presidential or congressional election years, for instance.

I have in this short resumé minimized the role economic, political, and even military factors ought to play in a truly comprehensive account of the origins of war. This has been done in order to bring to the fore the question of human sacrifice as a specifically religious aberration, not only in its internal dynamism but in its external trappings as well: flags and slogans soon become the icons and dogmas of blind belief. People who go to war are not homicidal maniacs; they are true believers in God, king, and country. It would put perhaps too sharp a point on it to call war the continuing failure of religious education, but certainly the issue does cry out for closer scrutiny from this angle. As noted, the muddiest material needing clarification centers on the scapegoat, the surrogate victim, the king yearly slain, and so forth. These are religious manifestations. In his monstrous book *Violence and the Sacred*,[194] René Girard throws down the gauntlet: religion, he says, is just a euphemism for bloody human sacrifice. It is a challenge that must be taken up, and the sooner the better.[195] Much as I am convinced that Girard has not the least inkling what religion is all about, I am also persuaded that he may be saying a great deal about the origins of war and rationales for collective human sacrifice.

The institution of "divine" kingship generally involves an identification of the king (that is, the state) with the reigning God, as either incarnation or representative. Here we cannot but remark the continuing collaboration of temple and citadel, church and state.[196] The sovereign arrogates to himself the divine power over Life and Death, and to do so requires the sanction of the priesthood. Only such religiously "approved" sovereignties could hierarchically organize entire societies, institute divisions of labor for construction or destruction, maintain centralized control, and in effect turn loose-knit neolithic village communities into the megamachine of civilization Mumford has so ably dissected: "Kingship is literally a man-eating device."[197]

From the point of view of the history of religions, it is also noteworthy that transcendence as, so to speak, a vertical differentiation—putting God over humans, and humans over nature, and one human being over all the others—quite probably preceded all the horizontal differentiation of human relationships into labor gangs and competing armies, each, be it noted, a kind of infernal machine composed of standardized, interchangeable, and disposable human components.

The transition from putatively divine sovereignty to the modern secular state came directly after the French Revolution, in 1789 to be exact. Sovereignty and all it implies—the right to kill to maintain social control and the hierarchical organization of society to this end—was severed from kingship less than two hundred years ago, but maintained all its perogatives in the new ideal of "national" sovereignty adopted by the republic. And just to bring us up to date, a minister in Texas whose flock makes nuclear weapons at the Pantex plant not long ago informed us on national television that Jesus would endorse a first strike against the "Godless" Russians. And so on some fronts the old collusion between temple and citadel continues seemingly unabated.[198] Sovereignty is, after all, an idea—or should we say a deeply ingrained social fiction?—a kind of cultural vampire which exists first and most formidably in the mind, and can be called to reckoning and (let us pray) exorcised only there.

The banality of modern evil, though much discussed,[199] is still by and large unseen where it dwells in our midst. We still expect evildoers to wear horns, and evil actions to be in some way grotesque. In fact it is precisely the blandness of modern horrors that so often keeps us from seeing them for what they are. An Eichmann talks like any other routinized bureaucrat, a pound of plutonium looks like harmless grey dust, a nuclear weapons plant looks like any other factory, an MX missile appears like any other item on the defense budget, and the people who design and build weapons a thousand times more lethal than the Nazi death camps consider themselves upright and patriotic citizens, decent family men who wear ties and brush their teeth and mow their lawns and go to church on Sunday. The "good German" mentality has lately become a focal point in discussions about the preparations for nuclear megadeath going on in seemingly innocuous ways all around us.[200] Paying income taxes to the United States government may be considered good citizenship, but when that government spends about 29 cents of every dollar collected on preparations for war, it is also complicity in the mechanistic and nowadays automated rituals of a Death Cult which promises only planetary annihilation. Such questions of conscience and complicity cannot of course be resolved in the abstract, but surely they must be raised if alternatives to Armageddon are ever to emerge.

Who controls the controller?

There seems to be something very wrong with the way we human beings have gone about arranging our collective affairs. Most of our attempts to institutionalize the values we cherish seem to backfire. As Ivan Illich consistently and convincingly demonstrates, our institutions tend in the long run to subvert the very values they were instituted to preserve;[201]

so that now, ever so subtlely, our basic values have been replaced by ersatz replicas and a police state is too easily mistaken for safety, a diploma for an education, a doctor's prescription for health, and the social rigidity of an armed camp for peace.

Where is the basic flaw in the cultural institutions—particularly those comprising the sovereign state—we have instituted to preserve our very (individual and collective) lives? We are not going to have an answer to this question until we investigate it, and such an investigation amounts to what I have earlier taken the liberty of calling a cultural hermeneutics: we must undertake a concerted and thorough rereading of human culture from the ground up. African army ants are one of the very rare examples of species other than our own which make war on their own con-specifics. It is perhaps more than a coincidence that as human society comes more and more to resemble an immense, mechanized, computerized, bureaucratic anthill, we come closer and closer to the ultimately self-destructive tendencies to which these pitiful ants fall prey. Man, the most dangerous species, has now become the most imminently endangered species. The doomsday clock of the *Bulletin of the Atomic Scientists* today reads three minutes to midnight.

The crux of this cultural miasma may well be the prevalent inability to conceive of the constitutive tensions and polarities of human life in any but dialectical terms: us/them, either/or, win/lose, yes/no, profit/loss, thesis/antithesis, male/female, black/white, left/right, etc. Computers can't "think" in any other way; the circuit is either on or it is off. The System (for present purposes, the war system) has most succinctly been defined by R. Buckminster Fuller: "Divide and conquer—and to keep conquered, keep divided."[202] This is the basis of the rather rudimentary system of control that now directs the destiny of western civilization and is threatening to drag the rest of the peoples of the world, kicking and screaming, into the abyss of its own predictable demise. It was first clearly formulated by Aristotle as the "self-evident" principle of non-contradiction mentioned earlier, and put most effectively into practice in the military campaigns of his most famous student, Alexander. It is the so-called excluded middle: the relationship between any two people or peoples is hypostasized and frozen into dialectical postures. Today this System—manifest in the heteronomic pyramid of corporate, military, professional, and bureaucratic power, where each rank obeys its superior without question or qualm of conscience—intervenes in every human relationship and monopolizes the center of every human transaction. (And I for one am not convinced that the creation of a new professional class of mediators or arbitrators is going to do much more than compound the problem.) The hidden premise here is that you and I are incapable of resolving our differences without recourse to a third party who provides a permit, license, or judgment—or, in case of failure, a convenient scapegoat or victim.

* * *

To summarize: in order to preserve the power of the intermediary (priest/king, state/bureaucracy, manager/administrator, and perhaps even negotiator/arbitrator) the entire medium of culture is weighted to suppress that vital center of spontaneity at the core of all human(e) intercourse, maintaining instead a milieu of competition in which everybody is, in the final analysis, at odds with everybody else. It's such a familiar story that we need not spell out all the details. It is also obvious enough that this System of (in)human relations is today in a shambles. Economic stopgaps and piecemeal technical or political repairs can no longer keep the megamachine running. The real question, again, is whether or not we human beings are capable of surviving its collapse.

The Nucleus

God's eye art'ou,
   do not surrender perception.
      Ezra Pound[203]

At the heart of the many interconnected concerns we have been discussing lies the open question of conversion, and this on both the personal and communal levels. When several trillions of U.S. tax dollars are slated to go to the defense establishment over the next few years, any resolution of the nuclear predicament short of holocaust is going to have massive societal repercussions. How do we convert a culture which has for all practical purposes become a Death Cult to life-affirming and life-enhancing values? Nearly all of our civilizational and institutional structures are enmeshed and enmired in the accelerated rush to oblivion. If culture is only an objective edifice out there in the world somewhere, it would seem as if we mere mortals have but little leverage with which to effect change. By the same token, if all our religious and ethical values are construed to be only subjective chimera in this world of hard realities, then they—and we, as thinking, feeling, caring human beings—are in the long run irrelevant and incapable of taking charge of our own lives. The entire import of the present book tends to confirm that there is no such rift between our ultimate (that is to say, religious) values and their worldly (that is to say, cultural) correlates and manifestations. Death is the only absolute subject/object dichotomy. But we behave as if we believe otherwise. Like sleepwalkers over the edge of a chasm, we behave as if all of us do not bear within us and between us not only the burden of having constructed a civilizational dead end, but also the seeds of renewal. John Adams once characterized the American revolution in exactly these terms: "The revolution took place in the minds of the people."[204]

In the same vein, if there is to be a turn toward Life in our culture two centuries later, a conversion from weaponry to livingry, then as we have stressed, the personal nexus of this transformation will be paramount. If you change your mind, you change the world[205]—but only if the mind here transformed is not some abstract, bloodless "ghost in the machine" but the *nous*—the mind, heart, and soul all together—said to be transfigured in the Christian *metanoia*.[206] Remember that the sense of the verb *to mind* is to care, to pay attention. Minding is caring; and therefore, secondarily, watching, watching over and watching out. *Metanoia*: a change of *nous*, the change from mindless distraction to mindful attention. The word certainly means much more than the "repentance" of the streetcorner preacher, although that's part of it. What *metanoia* really implies is the ever-deepening conversion and transformation of human(e) life that is supposed to be the sum and substance of the Christian message. It is a profound anthropological mutation, yes, but mutation understood as radical change, that is, not as a freak occurrence but as a change that keeps on changing. What is needed today is not just the exchange of one status quo for another, but instead a *fluxus quo*, a norm of change, transformation, renewal. Thus the conversion required is not so much a change of allegiance or a commitment to something else—to some new political platform, for instance, or even another religion—but a conversion that leads us directly into whatever and whoever we already are, and re-commits us to our own most vital possibilities.

If religion and culture, God and the world, the sacred and the secular, are watertight compartments (or totally segregated academic departments), then both are doomed. Only if we begin to recognize the human person as the true mediator of the entire reality, only if the person is seen to stand at the nexus, at the crossroads (*symbolon*[207]) of all the dimensions of the mystery of Life, do we have any hope of redressing the grossly malignant tendencies of our moribund civilization. This book has been an attempt on a small scale to relocate the center, the nucleus of the many-faceted nuclear issue where it belongs, in the constitutive human(e) relation-ships—bonds, links, roots, branches, and connections—with Heaven, Earth, and everybody in between. The book has also had the pragmatic function of outlining some of the major topics and unanswered questions involved in taking the nuclear issue into the classroom, specifically within the context of religious studies and religious education. The suggestion has emerged that religious studies can begin to play its proper role as a living nucleus of all the interdisciplinary studies which have as their common aim the rescue of humankind from the infernal and all too imminent danger of perishing utterly. As teachers, scholars, and students of religion and culture, we might do worse than to recall Simone Weil's astute reflection on the true aim of all education:

> Although people seem to be unaware of it today, the development of the faculty of *attention* forms the real object and almost the sole interest of studies. Most school topics have a certain intrinsic interest as well,

but such an interest is secondary. All tasks that really call upon the power of attention are interesting for the same reason and to an almost equal degree... because all of them develop that faculty of attention which, directed to God, is the very substance of prayer.[208]

The complex of issues which today converge on the problem of nuclear weapons and human extinction have formed a kind of spiritual glaucoma, a spreading blind spot on the retina of western civilization and on the various specialized lenses of academia as well. Perhaps because these problems are humanly depressing to an unprecedented degree, most of us have for over four decades looked away, refusing to face the predicament point blank. We presently have no choice but to do so: undivided attention to the nuclear issue has survival value; ignorance and avoidance of the issue have none. Engaging this faculty of attention may in fact be the true counterweight to the demonic collective automatisms currently propelling us to race, not drift, "toward unparalleled catastrophe." Despite its devaluation these days, education of this sort stands out as the highest human priority. As R. Buckminster Fuller, a very practical man, once put it,

> Quite clearly our task is predominantly metaphysical, for it is how to get all of humanity to educate itself swiftly enough to generate spontaneous social behaviors that will avoid extinction.[209]

The crux, therefore, is awareness—of ourselves, of one another, of the physical world we inhabit and of the metaphysical principles by which this whole concatenation of relationships coheres. And awareness dawns with the discovery of otherness, of relatedness: "No otherness, no awareness."[210] The pall of oblivion today hanging like an invisible mushroom cloud over every human head can be dispelled only by cultivating the awareness that we are intrinsically connected to the very well-springs of reality. This cultivation of an integral awareness has traditionally been understood as the nurture of human spirituality, according to a given discipline (see "Metalogue"). There is no way out, no escape from the contemporary human predicament, but there may very well be a way in, an opening and a deepening: *the direct awareness of reality*. This means not only the awareness that a subject might have of an object,[211] but preeminently the awareness *of* reality,[212] that is, the awareness that reality *has* , or rather *is*. In so many words, we human beings seem to be the locus of that integral awareness that reality has of itself. Or is that too strong? At least human awareness seems to be the intelligible face of the unfathomable, the face of the mystery turned toward us. The contemplatives of various traditions have always reminded us that the heart of reality is pure consciousness, or sheer consciousness, or even—since it is neither "consciousness-of"[213] nor "self-consciousness"[214]—a *mere* consciousness: the un-self-conscious transparency of Life at its source. Another word for this, in many ways a more abused word, is *love*: an awareness of the other (not an emotion, since it evokes the full gamut of emotions) so intense that it is not even aware of itself as awareness, much less of the other as an object. To experience this reality directly is also to discover oneself, not as a "self" at all, but as nothing more nor less than this awareness *that reality is*. It is therefore a wholly creative realization that in fact makes real the reality it real-izes.[215] It is the intensely personal and yet quite common experience of Death (disintegration) and Rebirth (integration), and it reverberates through every fiber of the real. At its core, reality would appear to be absolutely clear: awareness is that transpersonal center, the ultimate nucleus of the nuclear issue. And its nurture forms the crux of any study that would today presume to call itself religious.

Metalogue:
Spiritual Disciplines in the Nuclear Age

Earlier on, we discerned that religious peace has an interior as well as an exterior character, implying that there ought to be a personal equilibrium undergirding any semblance of tranquillity in the world at large. We spoke also, although perhaps too glibly, about cultivating an awareness at the basis of human community and creative communication. It is therefore clear that we are at some level or other already describing the character of spiritual disciplines in the nuclear age.

But how is this relevant to the concerns expressed in this book? At first go, spiritual disciplines seem a most unlikely and even unpopular topic. The phrase has fallen into disrepute. Even ignoring the current sadomasochistic connotations of "discipline," the word still carries undeniably authoritarian

overtones. Add "spiritual" to "discipline" and the modern
mind conjures up mainly images of ascetic excess: flagellation,
hair shirts, fasting, and many another implausible hardship.

Do we leave it at that? Does spiritual discipline mean simply
a form of bondage to one's own neurosis or somebody else's
psychosis, as the modern mentality would have us suppose?
Or is there more to it? And if we suspect there might be
a deeper sense hidden somewhere in the often labyrinthine
traditions of the world's religions, perhaps, how are we in the
apocalyptic climate of the nuclear age going to resuscitate this
more? Haven't times changed a good deal since most of the
religions formulated their understandings of spirituality?
Would such traditional understandings have any value in
today's hard-bitten, tough-minded world? Or, in the alternative,
are we capable of articulating entirely new visions of human
perfection and fulfillment, and thus entirely new regimens
for the attainment of such goals? We ought to look into all of
this, the traditional visions as well as their present and future
forms, a little more closely.

Terminology

Part of being human is to discover that we are not all that we could be, or should be, or would be. In every age and every culture, wherever we find human beings, we find them acutely aware of their own incompleteness, their limitations, their shortcomings. By the same token, wherever we find human beings, we find religion—precisely as this dimension of *more* to human life, this openness to potentiality, to being-in-the-making. Religions claim to be the ways to salvation, to completeness, to fully human (or superhuman) stature. Whether the way is envisioned as a positive construction built upon existing human qualities or as a negative deconstruction of the limitations of finite being, the human is an itinerant condition: being human is being on the way. . .to fullness or emptiness, to being or non-being, to the good life or the good death. Consequently, in religious traditions of the most diverse sorts we find various regimens, formal and informal, for bringing the human being to wholeness, however this integrity may be conceived. Etymologically, the holy is the "whole-ish," that which heals, which makes whole. Whole-making is the primary spiritual discipline—an integrity to be found in the constant balancing act that takes place day by day within each human being, between each of us and every other and, ultimately, between human consciousness and the whole of reality. Naturally (that is, culturally), there will be differences in emphasis between and among the religious traditions of humankind, just as there will always be differences according to personal temperament and idiosyncracy, but certain salient patterns do persist with enough regularity for us to propose a rough typology of spiritual disciplines world-wide. In order to understand the contemporary role of spiritual disciplines, we must first return to their traditional roots and try to understand the perennial themes they address.

Philology tells us that the word "discipline" (*disciplina*) is related to the Latin *discipulus*: a disciple, a learner (from *discere*, to learn); but philology cannot tell us exactly who is learning what from whom. There seems always to be a kind of tension or polarity, commonly misunderstood, between self-discipline (an autonomic, autodidactic model) and discipline imposed by another (a heteronomic, master/disciple model). It is presumably from the latter paradigm that the narrower sense of discipline as penalty or penance, punishment or chastisement, is derived, but this is not our concern here. The two models are in fact complementary and, as often as not, turn into one another: the autonomous urge to perfect oneself, to complete oneself by taking on one discipline or another, may be precisely the impulse that leads one to seek out a master capable of bringing such an apprenticeship to fruition. Contrariwise, if for the neophyte the master serves to embody that completion, that wholeness to which the disciple aspires, then the true master is the one who finally disappears, who "gives way" and allows the disciple to search out and follow his or her own path. But such paths are notoriously arduous, and the new wholeness too often aborted before it is born. Too often the disciple stumbles, falters, and stops halfway, or else the master sets himself up as an idol and thereby blocks the entire assimilation.

In the 20th century, we are also sensitive to a range of issues that barely surfaced at all in the classical formulation of religious ideals. We detect, for example, a certain evolution of spiritual disciplines from a traditionally privative emphasis (denying or depriving oneself or others of certain things, actions or intentions) to what is today emerging as a more inclusivistic emphasis (attempting to embrace every element and aspect of human experience). Moreover, we shall have to discover whether we are talking about spiritual *disciplines* (that is, disciplining the unruly spirit), or *spiritual* disciplines (that is, the exigencies of the spirit, which is sheer freedom). In other words, is it our intention to discipline the spirit, or to become its disciples?

We have suggested that the impulse toward wholeness governs human spirituality altogether. If this be the case, then the traditional spiritual disciplines by and large present themselves as attempts either to establish or to restore balance in human life. The need for discipline arises not only from the constitutive incompleteness of the human condition, but also from the fact that human beings are complex creatures who tend often to develop lopsidedly, emphasizing one dimension of the personality to the exclusion of others. Viewed as correctives to imbalance, three kinds of spiritual disciplines seem to call for our attention. Needless to add, a mature spirituality would embody all three, each in due degree.

Disciplines
of the Will

Suffering may well be the most universal sacrament in and through which human beings experience their intrinsic limitations. In suffering, physical and psychological, we discover the radical impotence of our will to control our own being. Through suffering, we discover a reality—whether called God or the devil, Life or Death—greater than we are, a reality we cannot manipulate at the very moment we most wish to do so. The true source of suffering is not physical pain, but the anguishing discovery that there is not, in the final analysis, very much we can do about it. The meaning of suffering is not something else, but only this: that our own will is not omnipotent, that our lives are not entirely in our hands, that we are not ultimately in control.

As a religious *primordium*, this painful discipline of the will is prefigured in the often inadvertent initiation of the shaman. The proto-shaman finds him or herself carried off to the world below, the world of sickness and suffering, the abode of King Death himself. Here he or she is dismembered or succumbs to one or all of the illnesses or is reduced to the barest essentials of life, to blood or bones for example. In this harrowing process, he or she either dies or learns to shamanize, to put body and soul back together again. It is nothing less than an experience of Death and Rebirth, which may indeed be the most basic religious experience of humankind. The shaman returns to the human realm a self-healed healer, and is thus prepared to help others make the same perilous passage. It is worth noting, further, that in telling of their adventures in the underworld, the shamans also generally re-collect all the elements that go together to form the lived world of their people, so that here personal transformation and world renewal tend to coalesce.

In western spirituality, the *locus classicus* for this sort of experience is the Book of Job. Following the conventional Jewish legalism of the day, suffering came as a punishment for sin, a notion dangerously akin to magic, since observing the Law would then seem to compel God to act in a certain way. Nonetheless, Job feels wronged. He committed no sin, and yet he suffers. He dares to call the Lord before him, as if before a court of law, to justify himself. But the voice of the Lord from the whirlwind offers a proof by revelation, not litigation. The Lord reveals himself as naked existence, in all its unconquerable majesty and uncompromising reality. If, as Jacopone di Toda once wrote, "The rose has no 'why,' " then neither have its thorns. Job learns this the hard way, through a total humbling of the will. The Lord here reveals himself, as he did to Moses in Exodus, precisely as YHWH: I AM WHO

AM: sheer being, the transparency of Life at its very source.
And existence does not reason or argue or make bargains. It
just is, and this is how it is. . . The image of the burning bush
from Exodus perfectly captures the sense of a creaturely
existence to which pain is intrinsic. The bush—the very Tree
of Life itself—burns, but is not consumed. Human life has
constantly to bear the burden of suffering, but it lives. And
not by my will, Job might have added, but by thine.

In eastern spirituality, the clearest classical approach to this
problematic would be the Four Noble Truths of the Buddha,
which read more or less as follows, and require practically no
comment:

1) The world is full of suffering.

2) The cause of suffering is desire.

3) Remove desire, and suffering will be ended.

4) In order to remove desire and end suffering, one must
follow the Eightfold Path: right view, right thought, right
speech, right behavior, right livelihood, right effort, right
mindfulness, and right concentration.

In all of the above instances, we are witnessing the power
of suffering itself to elicit a transformation at the very basis of
human being. Such transformative power might also in some
measure account for the common tendency to go to extremes
in this direction. Disciplines of the body—mortifications,
"little deaths"—are often most at home in traditions which
cultivate a thorough disdain, even disgust, for the body. In
our terms, these are attempts to subjugate the body's will or
natural inclinations to that of the spirit, here conceived as
separate, superior, and presumably insensitive to pain.

There has also often been what we today might term an
excessive emphasis on obedience in classical monastic formation,
both western and eastern. There is however a strong point to
such practices, too often lost on the modern mentality. It is

not the particular action demanded of the novice or disciple by the abbot or the master that is crucial, but the curbing of the will, the adherence to a higher authority. The focus is on the power of obedience itself, not its outcome or rationale. The abuses are well known, but the lesson itself—*ob-audire*, to hear the call of a higher will—is often missed.

The lesson learned from both the universal human experience of pain and these deliberate exercises in restraint would be simply that one's own will is weak, ineffective, and incomplete unless or until it is attuned to the primal Will that calls all things to be themselves and to transcend themselves. Only thus chastened is the will purified, healed, made whole and made free.

Disciplines of the Heart

Human life is built on, in, and through relationships. Nobody is or ever will be totally self-sufficient. We humans need each other even to be, and just as surely to become whoever we are. We need not only to love, but to be open to love. We are each incomplete without others. And through human love there breathes a breath of the Love that animates the universe, the love as Dante put it "that moves the sun and the other stars." The most widely practiced spiritual disciplines all involve the mystique of love; the devotional dimension always has been and today remains the most popular religiosity.

It is difficult to know where to begin. The devotional cult of Krishna in India or of Kuanon in China might provide exciting and exotic glimpses into the disciplines of the heart. But we need not go so far afield. In the western religious traditions, the most influential celebration of this intimate and intrinsic relativity we humans share with one another is the Song of Songs in the Old Testament. Here is the fountainhead from which Jewish, Christian, and Islamic traditions alike have drawn inspiration in matters of the heart, and of the soul.

The themes of this simple love song are universal, and the form of the lyrical dialogue between bride and bridegroom extraordinarily durable. The secret of its plasticity and adaptability is in this simplicity. The poem is constitutively susceptible to a variety of interpretations, and all the readings are valid, provided they leave room for the many other possible readings. Using only a conventional schema of interpretation like the medieval four levels—literal, figurative, allegorical, and anagogical—we shall shortly discern a dazzling spectrum of ways to approach the text. And when all is said and done, are not all disciplines of the heart intensely personal and unique and unrepeatable? The psychophysical chemistry of love is always going to be a little different for each one of us. That's the enchantment of it. For present purposes, it should be enough to open up each of the four classical approaches to the Song of Songs, and then try to put them back together again.

Literal

One's first and most immediate impression is that the Song of Songs is a very sensual celebration of earthly delights, focused specifically on the love shared by Solomon and the lovely Ethiopian (although the piece was probably composed much later, in the post-exilic period, 3rd century B.C.E.).

Note the particularity of detail: the cedars of Lebanon, the lilies, the flocks, the myrrh, balsam and honey. One can almost recognize the place. And there is no getting around the unabashed sexual metaphors:

> *Bridegroom:*
> She is a garden enclosed,
> my sister, my promised bride...
> *Bride:*
> Let my Beloved come into his garden,
> let him taste its rarest fruits.
> *Bridegroom:*
> I come into my garden,
> my sister, my promised bride...

This literal, physical character of the love song is understandably often ignored or downplayed in theological interpretations, but it is undeniably a prominent aspect of both the poem and the normal human experience of love. Nonetheless, one has an inkling that there may be more to it, that the beauty of the work is more than skin deep.

Figurative

It is possible to draw a moral from the story. One can see in the Song of Songs a figure for the relationship between any man and any woman. This figurative reading would emphasize the moral dimension of the poem, the prototype it sets for love and marriage, which may be why it is so commonly read at wedding ceremonies. The moral of the story is that love means total commitment. But the word *verse* itself means "a turning," and we quickly detect that what these verses give us is not a fixed doctrinal position but a moving picture, not the finale but the dance: the heady ecstasies and wrenching disappointments, all the sharp curves and sensuous contours, the passionate turnings toward and away from the beloved, all the ins and out and ups and downs of loving attention. Such an interpretation is attuned to the vibrant affective dynamisms of love, the fluctuations of yearning and desire fulfilled, the longing and racing of the heart. . .as he and she are now united, now divided; now sought, now found. Still and all, one suspects that even love and marriage may not yet be the whole story.

Allegorical

The Song of Songs has also been accepted as an elaborate allegory of the love story of God and Israel. This would in fact be the most common ecclesiastical interpretation of the poem. It is possible to find many correspondences between the changing fortunes of the bride and various sequences in Israel's history of conversion, falling away, hope, disillusion, and hope renewed. At one remove, the Christian Church later interprets itself as the bride and the poem prefigures the themes of union and distance, quest and fulfillment in the Church's relationship with Christ. At any rate, we are encouraged to chart the relationship of the "chosen people" to their God. But today's critical reader has difficulty accepting allegories; they seem arbitrary and contrived. If not a false reading (too many people have believed it for it to be wholly without value), it does at least seem a forced reading and thus impels one to look further.

Anagogical

It is as a mystical poem that the Song of Songs serves to chart the subtle ecstatic disciplines of the heart undertaken by many of the best-known visionaries in the Judaeo-Christian-Islamic tradition(s). Here the poem traces the relationship of the soul to God. Indeed, the mystical itinerary for the soul on its way to God is put into the lover/beloved format more often than into any other terms. Why should this be so? Probably first of all because there is often a great affective component in mystical experience, a tendency to favor the heart over the head, although the more astute contemplatives keep the balance and do not fall into mere sentimentality. Moreover, from primordial times, the sacred marriage has been celebrated as the embrace of Father Sky and Mother Earth, the union of spirit and matter, light and darkness, and so forth, and in almost every case the intimacy between a man and a woman seems to be its normal and natural human expression. The profusion of splendid works celebrating the divine spouse cannot be justly summarized here, but those of St. Bernard, Dante, Richard of St. Victor, Teresa of Avila, and the Sufis Rumi, Omar Khayyam, and Ibn el'Arabi might be enough to give one some sense of the range of possible variations on this inexhaustible theme. And yet even this most exalted interpretation cannot claim to be the whole story, since it amounts to a one-way translation of the very physical imagery of the poem to the spiritual heights of ecstatic experience. *Anagogical* means "leading to higher things," but is there no way to bring this lofty vision back down to Earth?

***

There is a simple way to draw together all the interpretive threads we have spun out so far. *Love* is the living symbol which allows us to assimilate the Song of Songs integrally, as a single seamless fabric. Through our own experiences of love we can sense the presence and the interpenetration of lover and beloved at every level: within each of us, between each human being and all others, between humankind and God, between God and the creation. The Song of Songs invites us to explore the depths of all the relationships that make us human; more, it is the discovery of the Divine as sheer relationship, and it echoes in every hidden alcove of the real. The lovers are real, and to be celebrated—in the flesh and in every homely aspect of daily life. God and "his people" have historically enjoyed a special relationship. And surely in any mystical experience, the soul and God grow always and only in and through their responsiveness to one another. The key here is the simple depth of pure love, which animates the myriad dynamic polarities of human/divine/cosmic interdependence and intercourse. All the verses—turnings—turn again into a single song of love: the one verse of many verses, the uni-verse, that is, the Song of Songs.

Disciplines of the Mind

The human mind is finite. Concepts are de-finite. The human mind has perhaps most characteristically conceived God as That which surpasses all definitions: the in-finite.

Drawing from Plato, Proclus, and perhaps certain initiatory schools of the Near East, the works of the anonymous 6th-century monk who called himself Dionysus the Areopagite (or the Pseudo-Dionysus) set the standard for the most rigorous disciplines of the mind in both Latin and Byzantine Christianity. The concluding passage from his *Mystical Theology* most clearly illustrates the *via negativa*, the "way of negation" which is of primary concern to us here. After he dismisses out of hand

any sensory, material, emotional, or imaginative conception
of God, we read the following:

> Once more, ascending yet higher we maintain that It
> is not soul, or mind, or endowed with the faculty of
> imagination, conjecture, reason or understanding;
> nor is It any act of reason or understanding; nor can It
> be described by the reason or perceived by the under-
> standing, since It is not number, or order, or greatness,
> or littleness, or equality, or inequality, and since it is
> not immovable nor in motion, or at rest, and has no
> power, and is not power or light, and does not live,
> and is not life; nor is It personal essence, or eternity,
> or time; nor can It be grasped by the understanding,
> since It is not knowledge or truth; nor is It kingship or
> wisdom; nor is It one, nor is It unity, nor is It Godhead
> or Goodness, nor is It a Spirit, as we understand the
> term, since It is not Sonship or Fatherhood; nor is It
> any other thing such as we or any other being can have
> knowledge of; nor does It belong to the category of
> non-existence or to that of existence; nor do existent
> beings know It as it actually is, nor does It know them
> as they actually are; nor can the reason attain to It to
> name It or to know It; nor is It darkness, nor is It
> light, or error, or truth; nor can any affirmation or
> negation apply to It; for while applying affirmations
> or negations to those orders of beings that come next
> to It, we apply not unto It either affirmation or
> negation, inasmuch as It transcends all affirmation by
> being the perfect and unique Cause of all things, and
> transcends all negation by the pre-eminence of Its
> simple and absolute nature—free from every
> limitation and beyond them all. *(C. E. Rolt, trans.)*

What an extraordinary de-conceptualization of the Godhead this is! Intellectually, confronting this text is like trying to climb a tall, sheer wall of glass. . .and slipping. Dionysus makes no concessions; he refuses to offer even a single foothold for rational explanation. The reader is left with precisely *nothing*—nothing but the most strenuously safeguarded openness to the depths of the divine mystery. Here the role of the true intellectual is brought to the fore: the intellectual ever stands—properly as guardian, improperly as interloper—at the threshold to the divine mystery. This far you may go with your reasons and concepts and opinions and logic, but no farther. God is not a reasonable man, much as some would-be theologians may be tempted to conceive him in their own image and likeness.

Negative theology like that of Dionysus is not for everybody. It is not a crowd pleaser. This is the steepest path, the "ascending way" which opens up only for the few. And these few are most often carefully selected and meticulously trained. Here is where the master/disciple relationship reaches its apex. The discipline here is not only a negative confrontation with the limits of human cognition, it is an initiation into the clear awareness at the core of reality, whether this be called God, Brahman, consciousness, or emptiness. Some of its living symbols are the trinity (or "Thearchy" as Dionysus calls it) in Christianity, *advaita* or nondualism in Hinduism, and the radical interdependence implied by the Buddhist *pratītyasamutpāda*. Its simplest expression would be the famous Sanskrit *neti, neti*: Not this, not that. The *Upanishads* and Shankara still stand as the most rigorous exponents of the nondual path in Hinduism, while Buddhism offers a variety of "Great Middle Ways" ranging from the Madhyamika of Nagarjuna to Zen archery. In Christianity, the so-called Neoplatonic tradition sparked by the Areopagite continued to incandesce for a thousand years, adding to its roster such luminaries as Scotus Erigena, Albertus Magnus, Bonaventure, the Victorines, Meister Eckhart, John of the Cross, Nicholas of Cusa, the anonymous author of *The Cloud of Unknowing,* and others less well known.

To put it another way, the *via negativa* is by no means a dead end. It is a living tradition which has matured and adapted and continues to develop in many spheres today. It is not restricted to the monasteries and the hallowed groves of academe. It should be clear by now that "negative" theology negates only the limits our limited minds would impose on ultimate reality. Some values are nonnegotiable; some mysteries inviolable. Some things are sacred.

The Contemporary Transition

Disciplines undertaken for their own sake achieve nothing. The classical forms of spiritual discipline we have examined so far were all means to an end, and that end is the liberation of the entire human personality, not merely one aspect or another. Just as any discipline involves a basic acceptance of finitude, so every form of discipline also has its own intrinsic limits. Hence it is necessary to underscore again the need for a proper balance between and among the paths outlined above. Humbling the will without passionate commitment and intellectual engagement produces mere reticence. Devotional attachment with no discernment and no restraint ends in blind fanaticism. And intellectual rigor pursued without love or humility comes to no more than cleverness. An integral spirituality is required, but the simplicity implied here is a complex achievement.

The problem is aggravated by the troubled times in which we live. Old ways of life are swiftly passing into oblivion, and entire systems of value are being turned on their heads. Most of the traditional spiritualities were geared to individual perfection, while today we cannot help but feel that the entire world and all its people are also in need of redemption. After Hiroshima, we know very well that if humankind is to survive we shall have to find alternatives to head-on conflict — creative, constructive ways to deal with the irreducible diversity of the human family. Working for peace has become one of the prime spiritual imperatives of our time. We are also aware we have taken the Earth for granted far too long, and that our spiritualities would be lopsided if they were only, so to speak, spiritual.

Another way to make the same point would be to show that secularization does not have to mean profanation. The process of secularization which has been so comprehensively documented over the past 20 years implies rather that the secular world so long despised in many of the traditional spiritualities is also real, that God is not cut off from this world and, consequently, that to seek God does not necessarily imply abandoning the world or one's fellow human beings. Today it means engagement with, commitment to, and responsibility for the whole of reality.

Thus we are in the midst of a transition, indeed a virtual transformation, from the static and alienating idea of individual salvation alone to the dynamic and integrative understanding that the entire reality is always implicated in any authentic personal effort to attain salvation. This transition is still underway, the transformation has not yet been fully realized, but a mighty threshold has been crossed. Contemporary spirituality has not repudiated the traditional themes discussed earlier, but we have to reckon with a nearly total turnabout in emphasis.

Instead of toughening up the discipline here in this "vale of tears" and postponing the promised deliverance to some other world and other time, the spirituality emerging in our day insists that the total liberation of the human person is real and possible *now*, this very moment, and furthermore that it is possible not just for a select group or a single aspirant, but for *everybody*. Similarly, the world itself is not to be ignored or disdained, but rather to be delivered from its own bonds of greed, exploitation, pollution, tyranny, and war. All of this indicates a tremendous shift taking place in the very idea of spiritual disciplines—from a private affair between the soul and God to a corporate and, indeed, planetary vision of human, divine, and cosmic cooperation.

* * *

One could present the transition discussed above in a negative way, by showing that humankind has come up against three new sets of limits and is beginning to evolve disciplines to deal with each of them. These would be, first, certain fundamental limits to the Earth's ability to support organic life; second, the ceiling placed upon human destructiveness by the abrupt discovery that our species has the power to engineer its own extinction; and third, the limitations intrinsic to any human attempt to plan, program, or manipulate Life, God, or reality. Put a little more positively, these three crucial dimensions of contemporary spirituality could briefly be summarized under the headings below.

The Goddess Aroused

Today the Earth is in pain. She suffers mightily, but no longer stolidly, from countless scars and about a billion open wounds. She is hurt, and one might almost say angry. The Earth has reached the limits of her ability to tolerate abuse from the most ungrateful of her offspring, those parasitic humans who treat her like dirt. They throw their garbarge at her, so now she's beginning to throw it back. They fill the air with smoke from their factories and automobiles, so she lets them choke on their own noxious fumes. They smash up the nuclei of her inmost atoms with wanton abandon, but she carries that decay and disintegration right back into their own cells, so that there is now detectable radioactivity in every human mother's milk. This is the lesson she must teach all her progeny: whatever they do to the Earth, they do to themselves.

Human negligence has aroused the sleeping Goddess, but aroused her wrath in the process. She seems to have returned with a vengeance, like Kali devouring her young. Entire species are going extinct every day. Food shortages, pollution, even ozone depletion and the greenhouse effect are beginning to take their inexorable toll in every ecological niche on the globe. What with wars of all sorts chequering our history, it must be conceded that our human species appears to have done its damnedest to turn this garden planet into a fiery inferno. And one day soon, we may very well finish the job. Yet nature will see to it that we are engulfed in our own firestorms, or else that the food chain dies out from under us, or maybe that we dump enough chemicals to bring about our own genetic apocalypse. Whatever we dish out to her, she will return in kind.

As we stressed earlier, the Goddess is not only this Earth, but what the ancients used to call the spirit of the Earth, *anima mundi*, the soul of the world, the whole of nature. She is the Mother because she is the *matrix*, the very womb of Life—not matter but that which "matters," the complex of metaphysical principles which materialize as the physical cosmos. The visible universe is but her raiment; the diaphanous jeweled gown—or is it a veil? The Goddess had until this tumultuous 20th century long been eclipsed in the West, seeming to capitulate at every turn to the mechanistic analyses and rational explanations of positivistic science. Today, as physics penetrates to the energy configurations underlying matter, as biology begins to read the genetic codex, and ecology to reassess all our relations with the material world, today we are sensing again the numinous presence of the Goddess—by whatever name: divine immanence, or cosmic intelligence, if you like, or just design integrity—on every front. In religious life as well, she has entered through many doors; through the spirituality of the Earth, of course, and environmental sensitivities of all sorts, but also through the alternative spiritualities emerging from the women's movement both inside and outside the established churches.

The question is still wide open whether the ultimate symbol here will be a revamped version of the Mediterranean Great Mother, or perhaps the White Tara of Tibet, or a more full-bodied appreciation of Mary in the historical traditions, or else the ancient and perennial *anima mundi* arrayed in the synergies and symbioses of holistic science or, more likely, a refreshing combination of all the above. The important thing to bear in mind is this renewed attunement to the cosmos as living organism, and the rediscovery of her most basic law: reciprocity. Why doesn't organic life on Earth just disintegrate? What holds entropy at bay? As we have observed all along, the world is held together not primarily by the cause-and-effect mechanisms detected by empirical science (mainly forces which increase entropy, randomness and disorder), but by a remarkably resilient fabric of connections, a wholly acausal matrix of minutely articulated relationships which cannot be ignored, or long neglected, or violated with impunity. And the most primordial of those relationships is that between Heaven and Earth, Father and Mother, creator and creation. They are not the same, obviously, nor are they antagonists. They are lovers, or at least at their best as lovers. However much the resurgence of Goddess religion may presently seem to be challenging the perogatives of the transcendent God in the highly patriarchal Jewish, Christian, and Islamic traditions, there is a wealth of precedent in all these traditions, as well as outside of them, for the sacred marriage of Heaven and Earth, *theos* and *cosmos*. Clearly God is not God for God alone, but for all creation. And the Goddess, once aroused, can be irresistible.

Human Interdependence

Our solidarity with the Earth and all her processes is
recapitulated and reinforced by our total interdependence
with one another. Ever since the summer of 1945, humankind
has had to live with the terrible knowledge that it is capable
of committing species-specific suicide. As we have more than
once emphasized, this looming problem of human extinction
is not going to go away. It will endure precisely as long as
human beings endure. The nuclear issue is generally treated
by the experts as a question of political strategy or military
tactics. As we have seen, the dilemma cuts much more deeply
than that into the very fabric of human life. The entire thrust

of this book has been to demonstrate that the true nucleus of the nuclear issue is how we human beings choose to relate to one another, constructively or destructively. The question of human relativity or interdependence comes down to this: We must learn in a single generation how to deal positively with conflict, or else we are all doomed, Q.E.D. If we do learn to make peace and not war, the achievement of this spiritual, moral, and physical discipline will be counted the most important legacy our generation leaves to those that follow. If we fail in this discipline, then we shall assuredly leave no legacy at all—for there will be no human beings alive to receive it.

What is needed is not a new law or a new superpower to enforce it, but a recognition of the law of interdependence which already holds sway over the destiny of humankind. Once this law—whether it is called reciprocity, relativity, equity, or justice—is acknowledged, certain activities are seen to be by their very nature outlaw activities. Most prominent among these would be the manufacture of any weapon which threatens to destroy the entire multidimensional web of life on this planet. Be it noted: this law of interdependence is in force whether it is ignored or not; its punishments are immediate and inescapable, and there is no higher court of appeal. If we human beings decide to destroy all life on Earth, nobody is going to slap our wrists and give us another chance. Were human interdependence to be spelled out in all its depth and power, it might take shape as something like the following "Declaration of Interdependence." Of course no single human being has the right to compose such a document, which would have to be ratified by humankind as a whole. But it may at least be permissible to submit a rough draft:

Declaration of Interdependence

We the free beings of biosphere Earth,
in order simply to survive, do hereby
affirm, ordain, and declare our total
and constitutive interdependence—
with one another, with all creatures
of the air, the waters, and the earth,
and with the mystery of Life itself.
Thus do we consecrate ourselves
to the service of Life,
and to one another.

By this declaration we do also sever all bonds,
whether legal, political, financial, or psychological,
which may yet bind us to mechanisms of planetary suicide.

There can be no lesser commitment. Only this immediate
and unconditional repudiation of the forces which presently
imperil the whole Earth can possibly forestall annihilation.
Recognizing and embracing our complete interdependence,
we hereby and for all time formally declare:
There is no place among us for instruments of genocide.
Be they nuclear, chemical, or biological, they are anathema.
Henceforward, the design, production, possession, and/or
deployment of any such instruments will be considered
crimes against humankind and against Life itself,
punishable by excommunication from humane society.
Use of such terrible weapons ineluctably exacts
its own terrible retribution.

Having no other home than this Earth, we cannot
and will not escape a shared destiny: Life, or death.
Having no other sane recourse, we choose Life.
Having no other allegiance than to the sublime
mystery of Life, we hold the following truths
to be obvious and inviolate:

> Life is the supreme value.
> Interdependence is the rule of Life.
> Love is the heart of this law.

Therefore,
be it resolved
by the free beings
of this biosphere Earth, that:

> This day we declare ourselves
> to be utterly, always, and only
> in the hands of one another
> —for better or for worse.

Mystery, Not Mastery

During this 20th century, the human mind has conquered the vast territories of the air and the deeps of the sea, extending the reach of human senses to the far planets and beyond. Age-old scourges like polio, smallpox, malaria, and diptheria have been dealt mortal blows. Technology has bent space and time into new and startling configurations. Jet airplanes, telecommunications, and interlocking computer networks have everywhere transformed human relationships. But there is, we cannot help but see, a dark side to all this progress, an ominous prospect of total destruction looming hand-in-hand with all the brighter trophies. The human mind has brought us to the brink of either well-being or nonbeing, utopia or oblivion. Today that very human mind faces what may well be its greatest challenge: it must conquer itself, it must discipline itself and, yes, restrain itself, or else cease to exist.

The human mind is a part of reality; it is not the whole. The danger that we are facing in the modern world is precisely that which arises when the part attempts to dominate the whole. The human mind has in our time laid claim to reduce, for all practical purposes, Life to logic, being to thinking and, ultimately, all the values of creation to the reasoning reason, the *ratio*, rationality. We too easily tend to ignore or not even to notice anything that is not rationally explicable, or quantifiable, or statistically predictable. Little by little we have tried to co-opt reality into this calculating frame of mind, to plan, maneuver, and manipulate Life right down to the tiniest atomic and genetic details. In a word, we have single-mindedly tried to box life into what we suppose are knowable parameters, partly perhaps to give ourselves a spurious sense of certainty or security in what is really a most insecure age.

But Life is full of surprises. Reality simply resists being reduced to any one of its dimensions. The attempt to manipulate Life to logical ends almost inevitably backfires for much the same reason that despite travel agents, plane schedules, and hotel reservations, one's vacations never quite turn out as anticipated. Everyone knows what befalls the best-laid plans of mice and men. Life can be planned or controlled only to a point; a point beyond which it asserts its own indefatigable creativity and spontaneously produces the unexpected: wonders and marvels, of course, but also atomic bombs, genetic mutations, madness, and misadventures of all sorts. In this sense, Life—human life especially—is free. It is not bound to grow or develop according only to logical premises or logistical strategies. Logic is derived from Life, not the other way round; reflection is born of experience. We cannot turn this most basic relationship on its head, like so many terms in an equation, without courting disaster. But we can learn to live within these intrinsic constraints. If it is true that human consciousness is but a part of the whole, it is also surely a part that may part-icipate creatively.

We have to get over the presumption that the human mind is either an oracle or some sort of gigantic filing cabinet. Nobody can ever perfectly predict or reliably engineer the future. Nor can any one human being, culture, language, or religion claim access to the total range of past or present human experience. The mind is not primarily a container, but a tuner or transceiver. Authentic thinking might best be defined as "not knowing," not attempting to fit reality or Life or God into our prefabricated constructs, concepts, and opinions. If you already know, you are not thinking. It is rather sheer openness to the novelty and unpredictability of Life that allows us to attune ourselves most astutely to its vital possibilities. In short, the mind thrives in the uncanny atmosphere of mystery, and withers in a climate of certainty.

There is to Life a depth dimension, an abyss of mystery, which can in no way be manipulated by the mind. You may call this abyssal dimension God, or the unknowable, or freedom, or *Wakan Tanka*, the Great Mystery; the names are not ultimate. But you cannot ignore this dimension without short-circuiting Life at its source. This abyss is the guardian of creativity, it is the very heart of love, the wellspring of intuition, the pivot on which everything turns. And it is truly a mystery, impermeable to us. It is why th lightning strikes where it does, and where our thoughts come from, and where all our deepest feelings go. There is more to Life than meets the eye, or resounds in the ear, or enthralls the mind. And it is this "more to it" that makes Life worth living, is it not?

An Integral Spirituality?

All the aspects of contemporary spirituality we have so far examined do not yet quite add up to a coherent whole. The integral spirituality that will rise to the challenges of our day is not to be found in some abstract set of categories, ready-made and pre-packaged. On the contrary, spirituality is always and only found as it is lived. If the spirituality emerging in our day is not to be stillborn, it will have to be forged anew in the living crucible of the human person. In this sense, the true path to the goal is always unique and wholly unpredictable. It will have deep roots in the millennial religious experience of humankind, and it will branch out into every aspect of contemporary life, but in the final analysis the spirituality each of us discovers for himself or herself will be a new thing under the sun, a true creation fitted to the exigencies of our own lives. As we have insisted from the outset, the human person is the authentic carrier of the entire reality, and every one of us is an exception to the law of averages, a statistical freak, a wondrous and utterly singular occurence. So we must resist the temptation to generalize unduly.

This much, however, we may say: Life is the art of graceful change. What we are looking for in the waning years of the 20th century is not an antique, or a prematurely finished product, or a capricious digression into eccentricity. We are tending rather toward a journeyman spirituality, a spirituality appropriate to the itinerant human condition, to being-on-the-way. . . Human beings are neither the masters of reality nor its hapless thralls. We are partners with the whole of Life—God, humans, and universe altogether—and, ultimately, only that whole coheres. Neither whole nor part can truly be said to exist, let alone make much sense, without the other. The whole will fall apart without the partnership of its parts, obviously; but the part itself only coheres so long as it plays its part in the integrity of the entirety. The spirit of the whole abides in every part.

We have seen that spiritual disciplines exist not to put a straitjacket on that spirit, but to release it. Now the spirit, it is well known, blows where it listeth. The kingdom of the spirit is the realm of sheer freedom, the freedom of being to be all that it can be. If everybody were already a disciple of the spirit, so to speak, and fully prepared to appropriate that freedom, there would be no need for disciplines. As it is, we tend to move ponderously, cautiously, step by slow step, into our own creative freedom, just as dancers and artists become graceful only after long years of effort. Free spirits are made, not born. Theirs may well be the most difficult discipline of all: openness, readiness, alertness. . .a kind of ultimate candor in the face of one's own experience.

The mandate of the spirit is pure spontaneity. A total human spontaneity, therefore, would constitute a total or integral spirituality: it just happens. Yet it happens only in the present, never elsewhere or at any other time. And when it happens, it comes to pass as clearly and effortlessly as the wind in the trees, or starlight in a winter night sky.

###

Notes

1. T. S. Eliot, "The Hollow Men" (1930), *Collected Poems 1909-35*, New York (Harcourt, Brace & Co.) 1936.

2. See, e.g., Lewis Mumford, *The Myth of the Machine*, vol. 1: *Technics and Human Development* and vol. 2: *The Pentagon of Power*, New York (Harcourt, Brace, Jovanovich) 1967/70, with extensive annotated bibliographies.

3. See Shingo Shibata, "The Tasks of Philosophy and Social Science to Prevent Nuclear Extinction-Philosophy and Social Science for Survival," in his book *Tasks of Our Time*, vol. 1: *For Abolition of Nuclear Weaponry*, in Japanese, Tokyo, 1978. Partial English translation, "Religion and Nuclear Extermination," in *The Churchman*, Aug.-Sept. 1979, pp. 6-7. Cf. also Shibata's pioneering *hibakusha* studies, e.g., *Alice Herz als Denkerin un Friedenskämpferin*, Amsterdam (Grüner) 1977.

4. Helen Caldicott, Address to *Women's Party for Survival*, 1980, transcript by Earth Education International, Goleta, CA, p. 17.

5. John Ernest, "A Call to the Conscience of the University Community," in *The Center Magazine*, Jan.-Feb. 1982, pp. 2-12.

194   NUCLEUS

6. See Jim Castelli, *The Bishops and the Bomb*, New York (Doubleday/ Image) 1983, which includes the Bishops' Pastoral in its entirety, "The Challenge of Peace: God's Promise and our Response," pp. 185-283.

7. Cf., e.g.,"Dead Letter?-Pastoral Picture Not Bright," in *Nuclear Times*, April 1985; Ira Chernus & Edward T. Linenthal, "Teaching Religious Studies in the Nuclear Age," in the *Bulletin* of the Council for the Study of Religion, vol. 15, no. 5, December 1984, pp. 141-3.

8. Ezra Pound, Canto 96, *The Cantos*, New York (New Directions) 1970, p. 655.

9. Max Scheler, *Formalism in Ethics and Non-formal Ethics of Values*, Evanston, IL (Northwestern University Press) 1973, p. 371.

10. Mary Douglas, "Morality and Culture," in *Ethics*, vol. 93, July 1983, pp. 230-1.

11. R. Panikkar, "The 'Moral' of Myth and the Myth of Morals," chapter III of his *Myth, Faith and Hermeneutics*, Ramsey, NJ (Paulist) 1979, pp. 38, 45.

12. Elizabeth Kübler-Ross, et al., eds., *Death-The Final Stage of Growth*, Englewood Cliffs, NJ (Prentice-Hall/Spectrum) 1975, pp. 164 ff.

13. Eugene Lauer and Joel Mlecko, eds., *A Christian Understanding of the Human Person*, Ramsey, NJ (Paulist) 1982, p. 3.

14. Carl Gustav Jung, *The Undiscovered Self*, Boston, MA (Atlantic/Little, Brown) 1958, pp. 12-13.

15. Martin Heidegger, *What Is Called Thinking*, Fred D. Wieck and J. Glenn Gray, trans., New York (Harper & Row) 1968, pp. 5-6.

16. E. Pound, Canto 99, *The Cantos*, op. cit., p. 698.

17. R. Panikkar, "Colligite Fragmenta: For An Integration of Reality," in *From Alienation to At-Oneness*, F. Eigo, ed., Proceedings of the Theology Institute of Villanova University, New York (Villanova University Press) 1978, pp. 19-91. Quotation from p. 74.

18. Cf., e.g., Mircea Eliade, *The Sacred and the Profane*, Willard R. Trask, trans., New York (Harcourt, Brace & World) 1959; Peter Berger, *The Sacred Canopy*, New York (Doubleday) 1967; R. Panikkar, *Worship and Secular Man*, Maryknoll, NY (Orbis) 1973; Harvey Cox, *Religion in the Secular City*, New York (Simon & Schuster) 1984.

19. See Matthew Fox, *Original Blessing, A Primer in Creation Spirituality*, Santa Fe, NM (Bear & Co.) 1983.

20. R. Panikkar, *Myth, Faith and Hermeneutics*, op. cit., pp. 451-2.

21. Albert Einstein, *Ideas and Opinions*, New York (Dell) 1973.

22. Pierre Teilhard de Chardin, *The Phenomenon of Man*, New York (Harper & Row) 1965.

23. Harvey Cox, from his *Millennium*, cited by Mary Long, "Visions of a New Faith," in *Science Digest*, Nov. 1981, p. 39. Cf. also R. Panikkar, "The End of History-Three Kairological Moments in Human Time-Consciousness," in *Teilhard and the Unity of Knowledge*, Thomas F. King & James F. Salmon, eds., New York (Paulist) 1983, pp. 101-2.

24. Ibid., p. 106-7.

25. Cf. R. Panikkar, *Blessed Simplicity, The Monk As Universal Archetype*, Scott Thomas Eastham, ed., New York (Seabury) 1982, pp. 128-9.

26. Cf. Walter W. Skeat, *A Concise Etymological Dictionary of the English Language*, New York (Capricorn) 1963, pp. 229, 609: Middle English *hool*, Anglo-Saxon *hal* (hale, whole) y = holy. Cf. German and Dutch *heilig*, that which heals, that which is "whole-ish" is the holy.

27. See Thomas M. Berry, "The New Story," in *The Riverdale Papers*, vol. 4, Riverdale Center for Religious Research, Riverdale, NY, 10471; and John F. Haught, *The Cosmic Adventure-Science, Religion and the Quest for Purpose*, Ramsey, NJ (Paulist) 1984. Cf. also Brian Swimme, *The Universe Is a Green Dragon-A Cosmic Creation Story*, Santa Fe, NM (Bear & Co.) 1985.

28. Cf. R. Panikkar, "Verstehen als Ueberzeugtsein," in *Neue Anthropologie*, Hans-Georg Gadamer and Paul Vogler, eds., vol. 7: *Philosophische Anthropologie*, Stuttgart (Georg Thieme Verlag) 1974, pp. 132-167; and also R. Panikkar, *The Intrareligious Dialogue*, Ramsey, NJ (Paulist) 1978.

29. Carl Gustav Jung, et al., *Man and His Symbols*, London (Aldus Books) 1964, pp. 72-3. Cf. also Carl Gustav Jung, *The Undiscovered Self*, R. F. C. Hull, Trans., Boston (Atlantic/Little, Brown) 1957, for this motif of "the shadow" and the cold war.

30. Cf. Martin Heidegger, *Being and Time*, John Macquarrie & Edward Robinson, trans., New York (Harper & Row) 1962, Introduction, I, 4, "The Ontological Priority of the Question of Being," p. 32, and VI, "Care as the Being of Dasein," pp. 225-273.

31. Cf. Gabriel Marcel, *Homo Viator*, Paris (Aubier) 1944.

32. Cf. R. Panikkar, "Have Religions the Monopoly on *Religion*?," in *Journal of Ecumenical Studies*, vol. XI, no. 3, Summer 1974, pp. 515-7.

33. Mircea Eliade, Conversations with Claude-Henri Rocquet, significantly titled *Ordeal by Labyrinth*, Chicago, IL (University of Chicago Press) 1982, p. 27.

34. Cf. Arthur O. Lovejoy, *The Great Chain of Being*, Cambridge, MA (Harvard University Press) 1936, 1964.

35. Cf. W. W. Skeat, *Concise Etymological Dictionary of the English Language*, op. cit., p. 52: "*Bless*, to consecrate, &c. (E.) The original sense may have been 'to consecrate by blood,' i.e., either by sacrifice or by the sprinkling of blood, as the word can be clearly traced back to *blood*, M.E. *blessen*, A.S. *blētsian*, O. Northumb. *blēdsia, bloedsia* (Matt. xxv, 34; xxvi, 26), which can be explained from *blod*, blood, with the usual vowel change from *o* to *oe* or *ē*."

36. There should be more than passing interest in pursuing this concept of "holy war," which seems especially prominent in the monotheistic religions native to the volatile Mediterranean basin, i.e., the policy of *herem* (or the ban) in Old Testament Judaism (e.g., I Sam. 15, 1 Kings 20, Deut. 13:15, Josh. 6:21), *jihad* in Koranic Islam (e.g., Suras 15, 33, 61, etc.), and crusade in Medieval Christianity. Cf. James A. Aho, *Religious Mythology and the Art of War*, Westport, CT (The Greenwood Press) 1981, chapter 8, "The Transcendental Historical War Myth and Military Ethic," and chapter 9, "Divine Transcendence: The Ancient Hebraic Military Ethic."

37. Cf., e.g., Robert Hyer, ed., *Nuclear Disarmament. Key Statements of Popes, Bishops, Councils and Churches*, Ramsey, NJ (Paulist) 1982; *Cross Currents*, vol. XXXII, no. 1, Spring 1982, entire issue entitled "No to Nuclear War"; R. Mahoney, "The Catholic Conscience and Nuclear War," in *Commonweal*, CIX, 12 March 1982, p. 138; K. Hormann, *Peace and Modern War in the Judgment of the Church*, Westminster, MD (The Newman Press) 1966; R. W. Tucker, *Just War and Vatican Council II: A Critique*, New York (The Council on Religion and International Affairs) 1967.

38. Cf., e.g., T. A. Shannon, Ed., *War or Peace? The Search for New Answers*, Maryknoll, NY (Orbis) 1980, Part I, "Just War Theory"; Robert M. Friday, "Nuclear War: The Bishops and Personal Conscience," in *The Living Light*, vol. 20, no. 1, Oct. 1983, pp. 24 ff.

39. See Knut W. Ruyter, "Pacifism and Military Service in the Early Church," *Cross Currents*, vol. XXXII, no. 1, pp. 54-70.

40. Cf. The Bishops' Pastoral, "The Challenge of Peace: God's Promise and Our Response," I.C., 'The Moral Choices for the Kingdom,' 1. The Value of Non-violence, in J. Castelli, *The Bishops and the Bomb*, op. cit., pp. 223ff.

41. See Luke 21:19; and Thomas Aquinas on "patience," *Summa Theologica* II-II, q. 136, a. 4. Cf. also R. Panikkar, *Myth, Faith and Hermeneutics*, op. cit., II. "Tolerance, Ideology and Myth," 3. 'The Four Moments of Tolerance,' d. Mystical, pp. 23-4, 35 (n. 9), and ibid., p.20: "The tolerance you have is directly proportional to the myth you live and inversely proportional to the ideology you follow."

42. Cf. R. Panikkar's approach to pluralism in "The Myth of Pluralism: The Tower of Babel-A Meditation on Nonviolence," *Cross Currents*, vol. XXIX, no. 2, Summer 1979, pp. 197-230.

43. Cf. Gregory Bateson, *Mind and Nature-A Necessary Unity*, New York (Dutton) 1979, e.g., in the Glossary, p. 227: "*Co-evolution.* A stochastic system of evolutionary change in which two or more species interact in such a way that changes in species A set the stage for the natural selection of changes in species B. Later changes in species B, in turn, set the stage for the selecting of more similar changes in species A."

44. Donald Keys, *To Arrest Nuclear War*, Menlo Park, CA (Planetary Citizens) 1982, pp. 3ff.

45. Cited in ibid., p. 4. Cf. also Donald Keys & E. Laszlo, *Disarmament, The Human Factor*, New York (Pergamon Press) 1981, available from Planetary Citizens, 777 UN Plaza, New York, NY 10017.

46. Cf., e.g., W. Taylor Stevenson, *History as Myth*, New York (Seabury) 1965, and R. Panikkar, "The End of History," op. cit, pp. 83-141.

47. Cf. George Steiner, *After Babel-Aspects of Language and Translation*, London (Oxford University Press) 1975, chapter 1, "Understanding as Translation,"
pp. 1-48.

48. William Carlos Williams, *Spring and All* (1923), Frontier Press, 1970, p. 80.

49. Cf. R. Panikkar, *Myth, Faith and Hermeneutics*, op. cit., chapter IV, "The Myth of Prajāpati. The Originating Fault or Creative Immolation," and chapter V, "Śunahśepa. A Myth of the Human Condition," especially 'The Sacrifice,' pp. 125-7.

50. Cf. R. Panikkar, "Man as a Ritual Being," in *Chicago Studies*, vol. XVI, no. 1, pp. 5-28, especially p. 18.

51. E. Pound, Canto CXII, *The Cantos*, op. cit., p. 784.

52. Gregory Bateson, Open Letter to the University of California Board of Regents, January 1979. Cf. also G. Bateson, "The Pattern of an Armaments Race," *Bulletin of the Atomic Scientists*, 1946(!).

53. Psalm 45(46).

54. Lewis Mumford, *The Myth of the Machine*, op. cit., p. 232.

55. Arthur Waskow, "Mystery Not Mastery," in *Co-Evolution Quarterly*, no. 39, Fall 1983, pp. 43-6.

56. Cf. R. Panikkar, *Blessed Simplicity, The Monk as Universal Archetype*, op. cit., pp. 86-8, 106-8.

57. Cf. R. Panikkar, *The Vedic Experience*, Berkeley/Los Angeles (University of California Press) 1977, "The Encounter," pp. 746ff.

58. Cf. R. Panikkar, *The Trinity and the Religious Experience of Man*, New York/London (Orbis/Darton, Longman & Todd) 1973, for a lucid exposition of the Trinity as the key to the encounter of religions.

59. *Advaita* means "un-divided"; not dualism, but not monism either. Its simplest expression may be the Sanskrit *neti, neti*: literally, "not this, not this." Cf. Eliot Deutsch, *Advaita Vedanta, A Philosophical Reconstruction*, Honolulu (University Press of Hawaii) 1969; and R. Panikkar, "Ṛtatattva: A Preface to a Hindu-Christian Theology," in *Jeevadhara*, no. 49, Jan.-Feb. 1979, pp. 6-63.

60. Usually rendered "dependent co-origination" — i.e., interdependence, original relatedness. See Frederick J. Streng, *Emptiness. A Study in Religious Meaning*, Nashville/New York (Abingdon) 1967. Cf. also R, Panikkar, *El Silencio Del Dios*, Madrid (Guadiana) 1970, "Pratītyasamutpāda," pp. 94-109.

61. C. E. Rolt, trans., *Dionysus the Areopagite*, London (SPCK) 1920, "The Divine Names," XI, 2, p. 175.

62. Norman O. Brown, *Love's Body*, New York (Random House/Vintage) 1966, pp. 181-2. Cf. also Jim Garrison, *The Darkness of God. Theology after Hiroshima*, Grand Rapids (Eerdmans) 1982, for a conscientious attempt (by means of Jungian psychology and process theology) to synthesize the nuclear issue and traditional apocalyptic, e.g., invoking the wrath of God, Hiroshima as gateway to Christ crucified, etc.

63. Cf. Bernard McGinn, *Apocalyptic Spirituality*, Ramsey, NJ (Paulist) 1979, Norman Cohn, *The Pursuit of the Millennium*, London (Oxford University Press) 1961, and J. A. Aho, *Religious Mythology and the Art of War*, op. cit., for some of the prototypical eschatologies of "holy" warriors.

64. Matt. 18:20.

65. Major Clifford H. Douglas in *Economic Democracy*, London, 1920, cited by Hugh Kenner in *The Pound Era*, Berkeley/Los Angeles (University of California Press) 1971, p. 311: "What men gain by not being isolated from one another he was to call 'the increment of association.' Like the cultural inheritance, it has demonstrable economic worth."

66. R. Buckminster Fuller's preferred definition of *synergy*, drawn explicitly from the synergetic behavior of chemical compounds and metallurgical alloys. See R. Buckminster Fuller and E. J. Applewhite, *Synergetics-Explorations in the Geometry of Thinking*, and *Synergetics 2*, New York (Macmillan) 1975, 1979, § 100.00, "Synergy," pp. 3ff.

67. Lyall Watson, *Lifetide*, New York (Bantam) 1980, pp. 147-8. Cf. also Ken Keyes, Jr., *The Hundredth Monkey*, Coos Bay, OR (Vision Books) 1982, where the story is retold with reference to the nuclear issue.

68. "Theon gar esmen synergoi." (I Cor. 3:9)

69. Keith Jarrett, *Survivor's Suite*, ECM Records, 1978, cover note.

70. Cf. Lewis Mumford, *My Works and Days*, New York (Harcourt, Brace, Jovanovich) 1979, chapter 24, "ABC of Demoralization," pp. 436-461, which includes all the major pieces Mumford has written thematically on the nuclear issue — beginning 8 August 1945! No one was quicker to react to the atomic bomb with moral outrage, no one more perceptive in his predictions of the consequences of an unchecked nuclear arms race.

71. R. Panikkar, "The End of History," op. cit., loc. cit., p. 103.

72. Cf. e.g., A. Einstein, *Ideas and Opinions*, op. cit., Part V, "Contributions to Science," for his essays written for laymen, "What is the Theory of Relativity," "On the Theory of Relativity," "Physics and Reality," etc., pp. 213ff.

73. Niels Bohr, *Atomic Physics and the Description of Nature*, London (Cambridge University Press) 1934, p. 96. Cf. also Niels Bohr, *Atomic Physics and Human Knowledge*, London (Chapman & Hall/John Wiley & Sons) 1958.

74. Except for certain anomalies like the occasional quantum léap which so infuriated Schrödinger. Cf. Gerald Holton, "The Roots of Complementarity," in *Daedalus*, vol. 99, 1970, pp. 1015ff, reprinted in the *Eranos Jahrbucher*.

75. Aristotle, *Metaphysics*, Book XI (Kappa), 5, 1062a.

76. Cf. R. Panikkar, "Singularity and Individuality. The Double Principle of Individuation," in *Revue Internationale de Philosophie*, 111-112, 1975, fasc. 1-2, pp. 141-166, which notes that the principle of identity (A = A) has served Hinduism as the presiding pattern of intelligibility quite as thoroughly as the principle of non-contradiction (A $\neq$ B) has served the West. Proceeding according to differences, one ultimately arrives at a God who is wholly other; proceeding according to the principle of identity, one ultimately arrives at Brahman, the ground of being, the Self-identity common to every self (*ātman*).

77. Cf. R. Panikkar, "The Existential Phenomenology of Truth," in *Philosophisches Jahrbuch der Görres-Gesellschaft*, 64 Jahrgang 1956, pp. 27-54.

78. Cf. G. Holton, "The Roots of Complementarity," op. cit., loc. cit.; and also Werner Heisenberg, *The Physicist's Conception of Nature*, Westport, CT (The Greenwood Press) 1970; and *Physics and Philosophy, The Revolution in Modern Science*, New York (Harper & Row) 1958.

79. Cf., e.g., Ian Barbour, *Myths, Models and Paradigms*, New York (Harper & Row) 1974, especially chapter 5, "Complementary Models," pp. 71-91; Thomas Fawcett, *The Symbolic Language of Religion*, Minneapolis (Augsburg) 1971; Ken Wilber, *Quantum Questions-Mystical Writings of the Great Physicists*, Boulder/Boston (Shambhala) 1983.

80. Werner Heisenberg, "The Representation of Nature in Contemporary Physics," in *Symbolism in Religion and Literature*, Rollo May, ed., New York, 1960, p. 209.

81. See Robert Jungk, *Brighter Than a Thousand Suns-A Personal History of the Atomic Scientists*, James Cleugh, trans., New York (Harvest/HBJ) 1958, especially chapters 2 and 3, "The Beautiful Years (1923-1932)" and "Collision with Politics (1932-1933)," pp. 10-47. Originally published as *Heiler als Tausend Sonnen*, Bern (Alfred Scherz Verlag) 1956.

82. Ibid., p. 54.

83. Ibid., p. 50.

84. According to Hans Fischer-Barnicol of the Heidelberg Institute for Symbolic Research, who interviewed Heisenberg for the details of this agreement.

85. R. Jungk, *Brighter Than a Thousand Suns*, op. cit., p. 91 n.; cf. also ibid., pp. 101-4, for Heisenberg's version of the story. According to Malcolm MacPherson's recent *Time Bomb-Fermi, Heisenberg and the Race for the Atomic Bomb*, New York (Dutton) 1986, Heisenberg recognized the ethical dilemma posed by atomic research well before most of his emigré colleagues, and reflected during the course of the war: "But is the use of an atomic bomb, by which hundreds of thousands of civilians will be killed instantly, warrantable even in defense of a just cause? Can we really apply the old maxim that the ends sanctify the means? In other words, are we entitled to build atom bombs for a good cause but not for a bad one? And if we take that view . . . who decides which cause is good and which is bad?"

86. The famous letter signed by Einstein is reprinted in, e.g., R. Buckminster Fuller & Anwar Dil, *Humans in Universe*, New York/Amsterdam (Mouton) 1983, pp. 46-7.

87. Cf. R. Jungk, *Brighter Than a Thousand Suns*, op. cit., chapter 11, "Atomic Scientist versus Atomic Bomb (1944-45)," pp. 171ff.

88. Ibid., p. 191.

89. Ibid., p. 333; to a visitor, perhaps in an attempt to leave his Faustian bargain behind him.

90. See ibid., Chapters 7 through 12, for accounts of the Manhattan Project by those directly involved.

91. Kurt Vonnegut's novel *Slaugherhouse Five*, New York (Delacorte Press) 1969, includes a dramatic and impassioned recreation of the Allied firebombing of Dresden, artistic capital of an earlier Germany.

92. The Sanskrit used in *Bhagavad Gita*, Book XI, is actually *kala*: "I am become Time, destroyer of worlds." But *kala* is also associated with the planet Saturn and thus, secondarily, with Death. For the occasion of the Trinity Test, I submit that Oppenheimer's version is the more appropriate reading: "I am become Death."

93. Cf. Jon Else's film, *The Day After Trinity*, for this interview and many others with original Los Alamos participants, e.g., Hans Bethe, Frank Oppenheimer and Stan Ulam.

94. For the Franck Report, submitted to the U.S. Secretary of War in June 1945, see R. Jungk, *Brighter Than a Thousand Suns*, op. cit., p. 184, and Appendix B, entire text.

95. J. Robert Oppenheimer, "Physics and the Contemporary World," in *The Open Mind*, New York (Simon & Schuster) 1955, p. 88.

96. R. Jungk, *Brighter Than a Thousand Suns*, op. cit., pp. 207-8.

97. Dallas *Times Herald*, 14 April 1984. Cf. also the recent film, *Genbaku Shi: Killed by the Atomic Bomb*, available from Public Media Arts, Inc., Santa Fe, NM.

98. See Takeshi Araki (Mayor, Hiroshima) and Yoshitake Morotoni (Mayor, Nagasaki), *Appeal to the Secretary General of the United Nations*, October 1976, which includes all-too-graphic illustrations.

99. Cf. R. Jungk, *Brighter Than a Thousand Suns*, op. cit., chapter 18, "In the Sign of the MANIAC (1951-1955)," pp. 297ff.

100. Cf. B. M. Russett and B. G. Blair, eds., *Progress in Arms Control?*, New York (W. H. Freeman) 1979; and the Stockholm International Peace Research Institute (SIPRI) *Yearbooks* on World Armaments and Disarmament, published by MIT Press, Boston, MA.

101. Cf. L. Mumford, "ABC of Demoralization," *My Works and Days*, op. cit., loc. cit.

102. Cf. Jungk, *Brighter That a Thousand Suns*, op. cit., pp. 302-5.

103. Ibid., p. 305.

104. Daniel Ellsberg, *Nuclear Armaments*, Berkeley, CA (Conservation Press) 1980, p. 1.

105. C.G. Jung, *The Undiscovered Self*, op. cit., chapter VI, "Self Knowledge"; composite reflections on the "shadow" assembled from pp. 95-6, 99-100.

106. United Nations, *Comprehensive Study on Nuclear Weapons*, 1981, Study Session 1, Dept. of Political and Security Council Affairs, UN Center for Disarmament, Report of the Secretary General.

107. Cf. Albert Szent-Györgyi, *The Crazy Ape*, New York (Grosset & Dunlap) 1970, chapter XVI, "Science and Society," pp. 69-73, for a defense of basic research. Cf. also Huston Smith's excursus on the "epistemology of control" in his recent *Beyond the Post-Modern Mind*, New York (Crossroad) 1982, chapter 4, pp. 62-91; and Jacob Needleman, *Sin and Scientism*, Interview by Robert Briggs, San Francisco (Broadside Editions/Robert Briggs Associates) 1985.

108. Cf. the now classic work on paradigm-shift in science by Thomas S. Kuhn, *The Structure of Scientific Revolutions*, Chicago/London (University of Chicago Press) 1962, 1970, as well as some of the more recent contributions of I. Barbour, T. Fawcett, K. Wilber, et al., cited earlier.

109. T. Berry, *Riverdale Papers*, op. cit., vol. 5, chapter 1, "The New Story," p. 1.

110. Cited in R. Jungk, *Brighter Than a Thousand Suns*, op. cit., p. 341. Cf. also Pauli's collaboration with C.G. Jung, one fruit of which is the powerful and insightful essay, "The Influence of Archetypal Ideas in the Work of Johannes Kepler." Many recent works testify to this "explicit or implicit imperative," e.g., Fritjof Capra's *The Tao of Physics*, Boulder, CO (Shambhala) 1976; B. Swimme, *The Universe is a Green Dragon*, op. cit.; Gary Zukav, *The Dancing Wu Li Masters*, New York (Morrow) 1979; as well as lesser known works like Virginia Stem Owens, *And the Trees Clap Their Hands*, Grand Rapids, MI (Eerdmans) 1983; and Elizabeth D. Gray, *Green Paradise Lost*, Wellesley, MA (Roundtable Press) 1981.

111. Cf. Lewis Thomas, *Lives of a Cell*, New York (Viking) 1974; James Lovelock, *GAIA-A New Look at Life on Earth*, London/New York, 1972.

112. Plato, *Timaeus*, 30c, 33b.

113. For example, the Pythagoreans, Plato, Aristotle, Proclus, the Pseudo-Dionysus, Augustine, Origen, Scotus Erigena, Bonaventure, Richard of St. Victor, Dante, St. Thomas, Avicenna, Ibn el-Arabi, Nicolas of Cusa, to name only a prominent few. And in the early 17th century, Johannes Kepler was still trying to build a geometrical model of it.

114. R. Panikkar, "Colligite Fragmenta," op. cit., loc. cit., p. 44.

115. Jan Christian Smuts, *Holism and Evolution*, New York (Macmillan) 1926, chapter V, "General Concept of Holism," p. 99. Cf. also Ken Wilber, *The Holographic Paradigm and Other Paradoxes*, Boulder/Boston (Shambhala) 1982, for both presentation and commendable critique of the recent attempts of Karl Pribram and David Bohm to articulate a unifying paradigm. Wilber emphasizes that the muddy philosophical trap here can be an extreme monistic reaction to the (still dominant) dualistic paradigms of mechanistic science, a reaction which tends to ignore all hierarchy and level every distinction.

116. Arthur Koestler, *Janus-A Summing Up*, New York (Vintage/Random House) 1978, pp. 269-70.

117. R. B. Fuller & E. J. Applewhite, *Synergetics* and *Synergetics 2*, op. cit.

118. Syn-tropic, i.e., integrative, not entropic. Cf. A Koestler, *Janus-A Summing Up*, op. cit., chapter XII, "Physics and Metaphysics."

119. G. Bateson, *Mind and Nature: A Necessary Unity*, op. cit., p. 8.

120. Scott Thomas Eastham, *Paradise & Ezra Pound-The Poet As Shaman*, Lanham, MD (University Press of America) 1983, p. 3.

121. Jonathan Schell, *The Fate of the Earth*, New York (Knopf) 1982, part II, "The Second Death," pp. 99ff.

122. The most comprehensive survey of this material is to be found in the volume *Hiroshima and Nagasaki*, by the Committee for the Compilation of Materials on Damage Caused by the Atomic Bombs in Hiroshima and Nagasaki, originally published by the Cities of Hiroshima and Nagasaki, now available in translation by E. Ishikawa and D. L. Swain, New York (Basic Books-Harper/Colophon) 1981.

123. We do have battlefield tactical weapons with low yields that approximate the first atomic bombs, roughly 12.5 kilotons for Hiroshima and 18 kilotons for Nagasaki.

124. Winners of the 1985 Nobel Peace Prize, the Physicians for Social Responsibility not long ago changed their name to International Physicians for the Prevention of Nuclear War, Inc., 225 Longwood Ave., Room 200, Boston, MA 02115. Materials from Ground Zero can as of this writing be obtained from 806 15th St., NW, Washington, D.C. 20036.

125. For too long a while, the major government publication with these figures has been *The Effects of Nuclear War*, compiled in 1979 by an advisory panel chaired by David Saxon and available in Government Printing Office bookstores or from the U.S. Congress, Office of Technology Assessment, Washington, D.C. 20510.

126. Cf., e.g., the *Nuclear War Prevention Kit* put together by the Center for Defense Information, 303 Capitol Gallery West, 600 Maryland Ave., SW, Washington, D.C. 20024, which also publishes the *Defense Monitor* to update its very reliable figures regularly. Another good source, e.g., for tapes of the well-known documentary *The Last Epidemic*, is the Cambridge Forum, 3 Church Street, Cambridge, MA 02138.

127. Helen Caldicott, "A Physician's View of Nuclear War," Interview, Berkeley, CA (The Conservation Press) 1980, reprinted in the Sierra Club *Yodeler*, January 1980.

128. U.S. Senate estimate by the Office of Technology Assessment.

129. Cf., e.g., "Accidental Nuclear War: The Human Factor," *PSR Newsletter*, vol. III, no. 4, Winter 1982, pp. 1ff. The excellent *PSR Newsletter* is available from 639 Massachusetts Ave., Cambridge, MA 02139.

130. Cf., e.g., J. Tirman, ed., *The Fallacy of Star Wars: Why Space Weapons Can't Protect Us*, New York (Vintage) 1984; and Robert M. Bowman, *Star Wars-A Defense Expert's Case Against the Strategic Defense Initiative*, Los Angeles, CA (J. P. Tarcher) 1986.

131. Related by Daniel Ellsberg in his Spring 1979 lecture series at Stanford University.

132. Cf. Helen Caldicott, "The Medical Implications of Nuclear Weapons Production," Address, 1980, tapes and transcript by Earth Educational International, Goleta, CA. Cf. also the increasing number of studies of the "atomic veterans" deliberately exposed by the U.S. Army to dangerous levels of radiation in the fifties who are now suing the government in droves for their higher incidences of cancer, birth defects, etc.

133. John Goffman, Remarks to *President's Select Committee on Nuclear Waste Disposal*, San Francisco Hearing, 22 July 1979.

134. Cf. the pitifully meager section "Incalculable Effects" of OTA's *The Effects of Nuclear War*, op. cit., pp. 114-5; and also B. Bennett, "Fatality Uncertainties in Limited Nuclear War," a report prepared for the Air Force by RAND Corporation (R-2218-AF), November 1977.

135. The best account of this devastated period is still Johan Huizinga, *The Waning of the Middle Ages*, New York (Doubleday/Anchor) 1954. Cf. also Barbara Touchman's novel, *A Distant Mirror*, New York (Random House) 1978.

136. Cf. Daniel Ellsberg, "Ashes, Ashes, All Fall Down . . .," Remarks delivered 20 February (Ash Wednesday) 1980, University of California, Berkeley, CA.

137. Cf. *The Last Epidemic*, Cambridge Forum, op. cit.

138. See Paul Ehrlich and Carl Sagan, *The Cold and the Dark*, New York (W. W. Norton) 1984. Audiovisual presentations of the "nuclear winter" scenario available from Center on the Consequences of Nuclear War, 3244 Prospect Street, NW, Washington, D.C. 20007.

139. Cf. L. Mumford, "ABC of Demoralization," *My Works and Days*, op. cit, loc. cit.

140. It might help if countries like the U.S. and France, signatories of the Nonproliferation Treaty, would stop setting such bad examples — not only by the bombs they build, but by their export of nuclear reactors and fissile material to South Korea, the Philippines, etc.

141. R. Panikkar, "The End of History," op. cit., loc. cit., p. 110.

142. Ibid., p. 133, n. 111.

143. Cf., e.g., Susan George, *How the Other Half Dies. The Real Causes of World Hunger*, London (Penguin) 1976.

144. Garrett Hardin may be today's leading neo-Malthusian predator, and is in some part responsible for lending an aura of respectability to the cutthroat "lifeboat ethics" espoused by many "have" nations in an effort to disengage themselves from any responsibility toward "have-not" nations. Cf. G. Hardin, "The Tragedy of the Commons," *Science* 161, 1968, pp. 1243-48; and G. Hardin, *Promethean Ethics: Living with Death, Competition and Triage*, Seattle, WA (University of Washington Press) 1980. Hardin's logic is impeccable, but his assumptions are not. Once you assume an inherent scarcity of life-support on planet Earth (an assumption first formulated by Thomas R. Malthus, *An Essay on the Principle of Population* (7th edition, 1872), New York (A. M. Kelley) 1971), then logic will inevitably lead you to conclude that some people will "naturally" have to die in order that others may live. R. Buckminster Fuller, for one, disputes Malthus directly. In his *Critical Path*, New York (St. Martin's Press) 1981 and *Grunch of Giants*, New York (St. Martin's Press) 1983 and elsewhere, he consistently focuses on the capacity of the human mind to do "more and more with less and less" expenditures of resources, thus making possible a higher standard of living for everybody than anybody has heretofore enjoyed. Fuller's premise is backed up in considerable detail by the *World Game* studies on food, energy, housing, etc., available from Southern Illinois University, Carbondale, IL.

145. Martin Heidegger, *The End of Philosophy and the Task of Thinking*, Joan Stambaugh, trans., New York (Harper & Row) 1973, p. 109.

146. Giles Gunn, *The Interpretation of Otherness*, New York/London (Oxford University Press) 1979, pp. 27-8.

147. Ibid., Chapter 5, "American Literature and the Imagination of Otherness," pp. 175ff.

148. George Steiner, *After Babel*, op. cit., p. 473.

149. John Haines, *Living Off the Country: Essays on Poetry and Place*, 1984.

150. Jorge Luis Borges, "The End of the Duel," in *Dr. Brodie's Report*, New York (Dutton) 1972, pp. 43-50; originally *El informe de Brodie*, Buenos Aires (Emecé Editores, SA) 1970.

151. René Dubos, *Celebrations of Life*, New York (McGraw-Hill) 1981, chapter 4, pp. 131ff.

152. R. Panikkar, "Hermeneutic of Religious Freedom: Religion as Freedom," *Myth, Faith & Hermeneutics*, op. cit., p. 450. Cf. also R. Panikkar, "Die Zukunft kommt nicht später," in *Vom Sinn der Tradition*, L. Reinisch, ed., München (C. H. Beck) 1970, pp. 53-64.

153. Cf. U.S. Catholic Bishops Pastoral, "The Challenge of Peace," op. cit., loc. cit., chapters IA "Peace and the Kingdom"; IB "Kingdom and History"; and IC, "Moral Choices for the Kingdom."

154. Cf. R. Panikkar, "Philosophy as Life-Style," in *Philosophes critiques d'eux-mêmes*, A. Mercier, ed., Bern (Peter Lang for FISP) 1978, p. 197.

155. R. Panikkar, "Have Religions the Monopoly on *Religion*?," op. cit., loc. cit., p. 17.

156. Cf. R. B. Fuller, *Synergetics*, op. cit., §301.00 "Definition: Universe," pp. 81-2: "The Principle of Synergetic Advantage (§229.) requires that we return to the Universe as our starting point in all problem consideration. We assiduously avoid all the imposed disciplines of progressive specialization. We depend entirely on our innate facilities, the most important of which is our intuition, and test our progressive intuitions with experiments."

157. Cf., e.g., *Comprehensive Study on Nuclear Weapons* (United Nations Study Series 1, 1981), Department of Political and Security Council Affairs, UN Center for Disarmament, Report of the Secretary General, UN Publication Sales No. E.81.I.11; and also *Disarmament, A Periodic Review by the United Nations*, published quarterly by the Department for Disarmament Affairs, UN Sales Section, Room DC2-853, United Nations, New York, NY 10017.

158. The Freeze is properly a grassroots movement, and one should contact one's local group for speakers, etc. Local groups keep in touch with one another by networking at various levels, e.g., through the Nuclear Weapons Freeze Campaign, 220 I St., Ste. 130, Washington, D.C.

20002 and SANE, 5808 Greene St., Philadelphia, PA 19144. Cf. also the recent monthly publication, *The Nuclear Times,* which packs a lot of information into a concise format. As a handy means for keeping abreast of the political dimensions of the issue, it is available from Nuclear Times, Inc., Room 512, 298 Fifth Ave., New York, NY 10001.

159. Cf., e.g., Robert F. Drinan, *Beyond the Freeze,* New York (Seabury) 1983. Fr. Drinan's idea of a Catholic crusade to eliminate nuclear weapons may be an appealing one in certain quarters, but insufficiently universal to do the job. The question of where next to direct the considerable energies of the Freeze movement remains open.

160. The *Nuclear War Prevention Kit* mentioned above (n. 126) lists rental and sales costs, and provides the addresses of organizations to contact for over 20 films, some excellent, e.g., *The Arms Race and Us, The Atomic Cafe, Between Men, The Bomb, Dark Circle, The Day After Trinity, The Defense of the United States* (CBS series), *Last Epidemic, Nuclear Nightmare, War Without Winners, The War Game,* etc.

161. *Time,* 13 January 1986, p. 39.

162. Cf., e.g., R. Panikkar, *Blessed Simplicity,* op. cit.

163. Cf., e.g., *The Whole Again Resource Guide,* Tim Ryan and Rae Jappinen, eds., annual periodical published by Source Net, Santa Barbara, CA (distributed by Capra Press) 1982; *The Next Whole Earth Catalog,* New York (Random House) 1980, updated regularly by *Co-Evolution Quarterly* and its successor, *The Whole Earth Review,* 27 Gate Five Road, Sausalito, CA 94965.

164. Cf., e.g., *Spiritual Community Guide,* published each Fall by Spiritual Community Publications, Box 1080, San Rafael, CA 94902; cf. also the "Politics and Religion" issue of *Co-Evolution Quarterly,* no. 39, Fall 1983.

165. Cf., e.g., Elizabeth Mann Borghese, *The Drama of the Oceans,* New York (Harry N. Abrams, Inc.) 1975; Richard Falk and Saul Mendlovitz, *Regional Politics and World Order,* New York (W. H. Freeman & Co.) 1973; Gerald and Patricia Mische, *Toward a Human World Order-Beyond the National Security Straitjacket,* Ramsey, NJ (Paulist) 1977; G. Clark and L. Sohn, *Introduction to World Peace through World Law,* Chicago (World Without War Publications) 1973.

166. Donald Keys, *Earth at Omega,* New York (Branden) 1985, p. 62.

167. Cf., e.g., R. Buckminster Fuller, *Operating Manual for Spaceship Earth,* Carbondale, IL (Southern Illinois University Press) 1969; René Dubos and B. Ward, *Only One Earth-The Care and Maintenance of a Small*

Planet, New York (W. W. Norton) 1972; Michael Katz, William P. Marsh and Gail Gordon Thompson, eds., *Earth's Answer-Explorations of Planetary Culture at the Lindisfarne Conferences*, New York (Harper & Row) 1977.

168. José Argüelles, *The Transformative Vision*, Boulder, CO (Shambhala) 1975; cf. also José and Miriam Argüelles, *Mandala*, Boulder, CO (Shambhala) 1972.

169. The word is *mir*. It is probably worth repeating the often cited statistic that there are more teachers of English in Russia than there are students of Russian in America. We might at least begin to encourage parity in language studies (providing the potential to understand one another), instead of seeking parity only in numbers of weapons (providing only the potential to destroy one another). One sign of life on this front might be the recent efforts of Educators for Social Responsibility, 23 Garden Street, Cambridge, MA 02138, to sponsor U.S./Soviet teacher and student exchanges.

170. Cf. Gaston Bouthoul, *Huit Mille Traités de Paix*, Paris (Payot) 1948, and M-bow Amadou-Mahtar, Ed., *Peace on Earth: A Peace Anthology*, Paris (UNESCO) 1975.

171. Cf. R. Panikkar, "The Myth of Pluralism-The Tower of Babel," in *Cross Currents*, vol. XXIX, no. 2, Summer 1979, pp. 197-230.

172. R. Panikkar, "The End of History," op. cit., loc. cit., p. 44.

173. Cf. R. Panikkar, "Epoché in the Religious Encounter," Chapter III of *The Intrareligious Dialogue*, op. cit., pp. 39-52.

174. W. W. Skeat, *Etymological Dictionary of the English Language*, op. cit., p. 102.

175. This view of the commons is of course exactly the opposite of that espoused by Garrett Hardin in "Tragedy of the Commons," op. cit., loc. cit., and "Is Violence Natural?," in *Zygon*, vol. 18, no. 4, December 1983, pp. 405-413.

176. Cf. Ivan Illich, *Tools for Conviviality*, New York (Harper & Row) 1973; *Toward a History of Needs*, New York (Pantheon) 1978; and "Vernacular Values," in *Co-Evolution Quarterly*, no. 26, Summer 1980, pp. 22-49.

177. Cf. H. Luijpen, *Existential Phenomenology*, The Hague (Martinus Nijhof) 1972, which concludes with an elucidation of this insight.

178. One need only cite an example or two to make the point, e.g., the massive and comprehensive studies anthologized by Richard A. Falk and Samuel S. Kim, eds., *The War System: An Interdisciplinary Approach*, Westview Special Studies in Peace, Conflict and Conflict Resolution, Boulder, CO (Westview Press) 1980; and *Toward a Just World Order*, vol.1, *Studies on a Just World Order*, Boulder, CO (Westview Press) 1982.

179. Cf. J. Strachey, ed., *Standard Edition of the Complete Psychological Works of Sigmund Freud, vol. 22, Why War?*, London (Hogarth) 1933. Cf. also Erich Fromm, *The Anatomy of Human Destructiveness*, New York (Holt, Rinehart & Winston) 1973, "Anthropology," pp. 129ff.

180. Cf. G. Hardin, "Is Violence Natural?," op. cit., loc. cit.

181. Cf., e.g., Richard Leakey, *Origins*, New York (Dutton) 1977, which contests Raymond Dart's "killer ape" theory of human development, as well as Robert Ardrey's "territorial imperative," by observing from the fossil evidence three salient traits exhibited by the hominids who eventually became *homo sapiens sapiens*: 1) a home base, 2) shared meals, and 3) cooperative action.

182. Cf. Fred Bodsworth, *The Atonement of Ashley Morgan*, New York (Dodd, Mead & Co.) 1962, a novel which explores this issue; and, for another view, Konrad L. Lorenz, *L'Instinct dans le Comportment de l'Animal et de l'Homme*, Paris (Masson et Cie) 1956.

183. C.G. Jung, *The Undiscovered Self*, op. cit., p. 95.

184. E. Fromm, *The Anatomy of Human Destructiveness*, op. cit., p. 163.

185. Stanley Diamond, *In Search of the Primitive-A Critique of Civilization*, New Jersey (Transaction Books) 1974, p. 1.

186. L. Mumford, *The Myth of the Machine*, op. cit., vol. 1, *Technics and Human Development*, p. 164. This entire section on the ancient megamachines of Egypt and Mesopotamia is relevant, pp. 162ff. (Pace, E. Fromm, *The Anatomy of Human Destructiveness*, op. cit., pp. 161ff.)

187. R. Panikkar, *Blessed Simplicity*, op. cit., p. 66.

188. J. Schell, *The Fate of the Earth*, op. cit., part III, "The Choice," pp. 179ff. A point first enunciated, I believe, by Bertrand Russell in *The Bulletin of the Atomic Scientists*, October 1946.

189. An all too respectable mace, even today, as Lewis Mumford noted: "When the British Parliament is in session, a gigantic specimen lies on the Speaker's table," *Technics and Human Development*, op. cit., p. 172.

190. Cf. R. Panikkar, *Myth, Faith and Hermeneutics*, op. cit., chapter IV, "The Myth of Prajāpati. The Originating Fault or Creative Immolation," pp. 65-95; and chapter V, "Śunahśepa. A Myth of the Human Condition," 2/b/i, "The Sacrifice," pp. 125-7, for the primordial motif of the sacrifice underlying Indo-European religiousness.

191. Cf. Joseph Campbell, *Myths to Live By*, New York (Viking/Bantam) 1972, chapter IX, "Mythologies of War and Peace," especially pp. 177-8.

192. Cf. Mircea Eliade, *Patterns in Comparative Religion*, New York (Sheed & Ward) 1958, 130. Human Sacrifice, 131. Human Sacrifice Among the

Aztecs and Khonds, 132. Sacrifice and Regeneration, and 134. Seeds and the Dead, pp. 341-347, 349-352.

193. L. Mumford, *Technics and Human Development*, op. cit., p. 150.

194. René Girard, *Violence and the Sacred*, P. Gregory, trans., Baltimore/ London (Johns Hopkins University Press) 1977, originally published as *La Violence et le sacré*, Paris (Editions Bernard Grasset) 1972.

195. Girard's is a structural challenge to religion altogether that is not quite met by espousing violence as a means of liberation from oppression, or by reverting to just war rationales for bloodletting, such as those offered by R. Macafee Brown in *Religion and Violence*, Philadelphia (The Westminster Press) 1973.

196. Cf., again, L. Mumford, "Kings as Prime Movers," in *Technics and Human Development*, op. cit., pp. 162ff., as well as "The Return of the Sun God," in *The Pentagon of Power*, op. cit., pp. 28ff.

197. L. Mumford, *Technics and Human Development*, op. cit., p. 184. For further critique of state sovereignty, see Jacques Maritain, *Man and the State*, Chicago (University of Chicago Press) 1951, chapter II, "The Concept of Sovereignty," pp. 28-53: "The two concepts of Sovereignty and Absolutism have been forged on the same anvil. They must be scrapped together." (p. 53)

198. Of course the Peace Pastoral of the U.S. Catholic Conference of Bishops — "The Challenge of Peace: God's Promise and Our Response," op. cit., loc. cit. — has gone some distance toward removing the priestly sanction from the nuclear adventurism of the U.S. government. But is it too little, too late? Have we truly reckoned with the extent to which the Christian Church (both Latin and Byzantine) has served as the vehicle for the very idea as well as the reality of imperialism (via the Holy Roman Empire, caesaro-papism, etc.) from its classical Mediterranean sources and models to the contemporary world? This question is posed forthrightly in the article "Empire," in the 1958 Edition of the *Encyclopaedia Britannica*, vol. 8, pp. 402-410, by Sir Ernest Barker.

199. See Hannah Arendt, *Eichmann in Jerusalem-A Report on the Banality of Evil*, New York (Penguin) 1963.

200. Cf., e.g., Milton S. Mayer, *They Thought They Were Free: The Germans 1933-1945*, Chicago (University of Chicago Press) 1955; R. Hilberg, *The Destruction of the European Jews*, New York (Quadrangle) 1967; Stanley Milgrim, *Obedience to Authority*, New York (Harper & Row) 1974; R. V. Sampson, *The Psychology of Power*, New York (Vintage) 1968, and similar studies.

201. Cf., e.g., Ivan Illich, *Deschooling Society*, New York (Harper & Row) 1970, 1971; *Medical Nemesis*, New York (Random House) 1976; *Toward a History of Needs*, op. cit.

202. Cf. R. B. Fuller, *Critical Path*, op. cit., pp. xxviii, 61-2, 162, 196, 214-15.

203. E. Pound, Canto CXIII, *The Cantos*, op. cit., p. 790. An idea which may well have originated with Meister Eckhart.

204. Ibid., Canto XXXVI, p. 157, drawing from the Adams-Jefferson correspondence.

205. Cf. H. Kenner, *The Pound Era*, op. cit., quoting Buckminster Fuller, p. 162: "Heisenberg said that observation alters the phenomenon observed. T. S. Eliot said that studying history alters history. Ezra Pound said that thinking in general alters what is thought about. Pound's formulation is the most general, and I think it is also the earliest."

206. Cf. R. Panikkar, *Myth, Faith and Hermeneutics*, op. cit., p. 102: "Metanoia is the condition for entering the kingdom of heaven, change of *nous*, of mind, of direction." Cf. also Matt. 3:2.

207. The greek *symbolon* was a name for the crossroads, then for the way marker at the crossroads. *Sym-ballein* means "to throw" (*ballein*) "together" (*sym*), as wayfarers going their separate ways are thrown together at the crossroads, and perhaps also as a dish or a vase is "thrown" by centering it on a potter's wheel. It is the opposite of *dia-ballein*, the diabolic, a throwing apart.

208. Simone Weil, *Waiting for God*, New York (G. P. Putnam's Sons) 1951, Harper/Colophon Edition, pp. 105-6.

209. R. B. Fuller & E.J. Applewhite, *Synergetics*, op. cit., Introduction, "The Wellspring of Reality," p. xxviii.

210. R. B. Fuller & A. Dil, *Humans in Universe*, op. cit., p. 127.

211. Objective genitive. The medieval schoolmen would perhaps call this knowledge *ratio*, the analytical, discursive, objective faculty; the reasoning reason.

212. Subjective genitive. The medievals also always distinguished the discursive *ratio* from the intuitive understanding called *intellectus*. Here is St. Thomas in *Quaestiones Disputate de Veritate*, 15, 1: "Although the knowledge which is most characteristic of the human soul occurs in the mode of *ratio*, nevertheless there is in it a sort of participation in the simple knowledge (*intellectus*) which is proper to higher beings, of whom it is therefore said that they possess the faculty of spiritual vision."

213. The "consciousness-of (objects)" of primary interest to the natural sciences.

214. The "self-consciousness (of subjects)" of primary interest to the humanities.

215. Cf. Keiji Nishitani, *Religion and Nothingness*, Jan Van Bragt, trans., Berkeley/Los Angeles (University of California Press) 1982, pp. 5ff.

Bibliography

Works Cited

Aho, James A., *Religious Mythology and the Art of War*, Westport, CT (The Greenwood Press) 1981.

Amadou-Mahtar, M-bow, Ed., *Peace on Earth: A Peace Anthology*, Paris (UNESCO) 1975.

Aquinas, St. Thomas, *Summa Theologica*, vols. I-III, New York (Benziger Bros.) 1947.

Araki, Takeshi (Mayor, Hiroshima) and Morotoni, Yoshitake (Mayor, Nagasaki), *Appeal to the Secretary General of the United Nations*, October 1976.

Arendt, Hannah, *Eichmann in Jerusalem-A Report on the Banality of Evil*, New York (Penguin) 1963.

Argüelles, José, *The Transformative Vision*, Boulder, CO (Shambhala) 1975.

Argüelles, José and Miriam, *Mandala*, Boulder, CO (Shambhala) 1972.

Aristotle, *Metaphysics*, Richard Hope, trans., Ann Arbor (University of Michigan Press) 1952.

Barbour, Ian, *Myths, Models and Paradigms*, New York (Harper & Row) 1974.

Barker, Sir Ernest, "Empire," 1958 edition of the *Encyclopaedia Britannica*, vol. 8, pp. 402-410.

Bateson, Gregory, "Open Letter" to the University of California Board of Regents, January 1979.
"The Pattern of an Armaments Race," *Bulletin of the Atomic Scientists*, 1946.

Bennett, B., "Fatality Uncertainties in Limited Nuclear War," a report prepared for the Air Force by RAND Corporation (R-2218-AF), November 1977.

Berger, Peter, *The Sacred Canopy*, New York (Doubleday) 1967.

Berry, Thomas M., *The Riverdale Papers*, vol. 4, Riverdale, NY (Riverdale Center for Religious Research).

Bodsworth, Fred C., *The Atonement of Ashley Morgan*, New York (Dodd, Mead & Co.) 1962.

Bohr, Niels, *Atomic Physics and the Description of Nature*, London (Cambridge University Press) 1934.
Atomic Physics and Human Knowledge,
London (Chapman & Hall/John Wiley & Sons) 1958.

Borges, Jorge Luis, *Dr. Brodie's Report*, New York (Dutton) 1972.

Borghese, Elizabeth Mann, *The Drama of the Oceans*, New York (Harry N. Abrams, Inc.) 1975.

Bouthoul, Gaston, *Huit Mille Traités de Paix*, Paris (Payot) 1948.

Brand, Stuart, ed., *The Next Whole Earth Catalog*, New York (Random House) 1980.

Brown, Norman O., *Love's Body*, New York (Random House/Vintage) 1966.

Brown, Robert Macafee, *Religion and Violence*, Philadelphia, PA (The Westminster Press) 1973.

Caldicott, Helen, "A Physician's View of Nuclear War," Interview, Berkeley, CA (The Conservation Press) 1980, and the Sierra Club *Yodeler*, January 1980.
Address, "Women's Party for Survival," 1980.
Address, "Medical Implications of Nuclear Weapons Production," 1980.

Campbell, Joseph, *Myths to Live By*, New York (Viking/Bantam) 1972.

Capra, Fritjof, *The Tao of Physics*, Boulder, CO (Shambhala) 1976.

Castelli, Jim, *The Bishops and the Bomb*, New York (Doubleday/Image) 1983, including the Pastoral Letter "The Challenge of Peace: God's Promise and Our Response," pp. 185-283.

Chernus, Ira, and Linenthal, Edward T., "Teaching Religious Studies in the Nuclear Age," in the *Bulletin* of the Council for the Study of Religion, vol. 15, no. 5, December 1984, pp. 141-3.

Clark, G., and Sohn, L., *Introduction to World Peace through World Law*, Chicago (World Without War Publications) 1973.

Cohn, Norman, *The Pursuit of the Millennium*, London (Oxford University Press) 1961.

Cox, Harvey, *Religion in the Secular City*, New York (Simon & Schuster) 1984.

de Chardin, Pierre Teilhard, *The Phenomenon of Man*, New York (Harper & Row) 1965.

Deutsch, Eliot, *Advaita Vedanta, A Philosophical Reconstruction*, Honolulu (University Press of Hawaii) 1969.

Diamond, Stanley, *In Search of the Primitive-A Critique of Civilization*, New Jersey (Transaction Books) 1974.

Douglas, Clifford H., *Economic Democracy*, London, 1920.

Douglas, Mary, "Morality and Culture," *Ethics*, vol. 93, July 1983, pp. 230ff.

Drinan, Robert F., *Beyond the Freeze*, New York (Seabury) 1983.

Dubos, René, *Celebrations of Life*, New York (McGraw-Hill) 1981.

Dubos, René and Ward, B., *Only One Earth-The Care and Maintenance of a Small Planet*, New York (W. W. Norton) 1972.

Eastham, Scott Thomas, *Paradise & Ezra Pound-The Poet As Shaman*, Lanham, MD (University Press of America) 1983.

Ehrlich, Paul and Sagan, Carl, *The Cold and the Dark*, New York (W. W. Norton) 1984.

Einstein, Albert, *Ideas and Opinions*, New York (Dell) 1973.

Eliade, Mircea, *Patterns in Comparative Religion*, New York (Sheed & Ward) 1958.
The Sacred and the Profane, Willard R. Trask, trans., New York (Harcourt, Brace & World) 1959.
Ordeal by Labyrinth, Conversations with Claude-Henri Rocquet, Chicago (University of Chicago Press) 1982.

Eliot, Thomas Stearns, "The Hollow Men" (1930), *Collected Poems 1909-35*, New York (Harcourt, Brace & Co.) 1936.

Ellsberg, Daniel, *Nuclear Armaments*, Berkeley, CA (Conservation Press) 1980.
"Ashes, Ashes, All Fall Down...," Address, 20 February 1980, University of California, Berkeley, CA.
Stanford University Lecture Series, Spring 1979.

Ernest, John, "A Call to the Conscience of the University Community," *The Center Magazine*, Jan.-Feb. 1982, pp. 2-12.

Falk, Richard A., and Kim, Samuel S., eds., *The War System: An Interdisciplinary Approach*, Westview Special Studies in Peace, Conflict and Conflict Resolution, Boulder, CO (Westview Press) 1980. *Toward a Just World Order*, vol. 1, Studies on a Just World Order, Boulder, CO (Westview Press) 1982.

Falk, Richard A., and Mendlovitz, Saul, *Regional Politics and World Order*, New York (W.H. Freeman & Co.) 1979.

Fawcett, Thomas, *The Symbolic Language of Religion*, Minneapolis (Augsburg) 1971.

Fox, Matthew, *Original Blessing, A Primer in Creation Spirituality*, Santa Fe, NM (Bear & Co.) 1983.

Friday, Robert M., "Nuclear War: The Bishops and Personal Conscience," in *The Living Light*, vol. 20, no. 1, Oct. 1983, pp. 24 ff.

Fromm, Erich, *The Anatomy of Human Destructiveness*, New York (Holt, Rinehart & Winston) 1973.

Fuller, R. Buckminster, *Operating Manual for Spaceship Earth*, Carbondale, IL (Southern Illinois University Press) 1969. *Critical Path*, New York (St. Martin's Press) 1981. *Grunch of Giants*, New York (St. Martin's Press) 1983.

Fuller, R. Buckminster, and Applewhite, E.J., *Synergetics-Explorations in the Geometry of Thinking*, and *Synergetics 2*, New York (Macmillan) 1975, 1979.

Fuller, R. Buckminster, and Dil, Anwar, *Humans in Universe*, New York/Amsterdam (Mouton) 1983.

Garrison, Jim, *The Darkness of God. Theology after Hiroshima*, Grand Rapids (Eerdmans) 1982.

George, Susan, *How the Other Half Dies. The Real Causes of World Hunger*, London (Penguin) 1976.

Girard, René, *Violence and the Sacred*, Patrick Gregory, trans., Baltimore/London (Johns Hopkins University Press) 1977.

Goffman, John, Remarks to *President's Select Committee on Nuclear Waste Disposal*, San Francisco Hearing, 22 July 1979.

Gray, Elizabeth D., *Green Paradise Lost*, Wellesley, MA (Roundtable Press) 1981.

Gunn, Giles, *The Interpretation of Otherness*, New York/London (Oxford University Press) 1979.

Haines, John, *Living Off the Country: Essays on Poetry and Place*, 1984.

Hamilton, Edith, and Cairns, Huntington, eds., *Plato: The Collected Dialogues*, Princeton (Bollingen/Pantheon) 1961.

Hardin, Garrett, *Promethean Ethics: Living with Death, Competition and Triage*, Seattle, WA (University of Washington Press) 1980.
"The Tragedy of the Commons," *Science* 161, 1968, pp. 1243-48.
"Is Violence Natural?," *Zygon*, vol. 18, no. 4, December 1983, pp. 405-413.

Haught, John F., *The Cosmic Adventure-Science, Religion and the Quest for Purpose*, Ramsey, NJ (Paulist) 1984.

Hilberg, R., *The Destruction of the European Jews*, New York (Quadrangle) 1967.

Heidegger, Martin, *Being and Time*, J. Macquarrie & E. Robinson, trans., New York (Harper & Row) 1962.
What Is Called Thinking, Fred D. Wieck & J. Glenn Gray, trans., New York (Harper & Row) 1968.
The End of Philosophy and the Task of Thinking, Joan Stambaugh, trans., New York (Harper & Row) 1973.

Heisenberg, Werner, "The Representation of Nature in Contemporary Physics," in *Symbolism in Religion and Literature*, Rollo May, ed., New York, 1960.
The Physicist's Conception of Nature, Westwood, CT (The Greenwood Press) 1970.
Physics and Philosophy, The Revolution in Modern Science, New York (Harper & Row) 1971.

Holton, Gerald, "The Roots of Complementarity," in *Daedalus*, vol. 99, 1970, pp. 1015ff, reprinted in the *Eranos Jahrbucher*.

Hope, Richard, trans., *Aristotle-Metaphysics*, Ann Arbor (University of Michigan Press) 1952.

Hormann, K., *Peace and Modern War in the Judgment of the Church*, Westminster, MD (The Newman Press) 1966.

Huizinga, Johan, *The Waning of the Middle Ages*, New York (Doubleday/Anchor) 1954.

Hyer, Robert, ed., *Nuclear Disarmament. Key Statements of Popes, Bishops, Councils and Churches*, Ramsey, NJ (Paulist) 1982.

Illich, Ivan, *Deschooling Society*, New York (Harper & Row) 1970, 1971.
Tools for Conviviality, New York (Harper & Row) 1973.
Medical Nemesis, New York (Random House) 1976.

Toward a History of Needs, New York (Pantheon) 1978.
"Vernacular Values," *Co-Evolution Quarterly*, no. 26, Summer 1980, pp. 22-49.

Ishikawa, E., and Swain, D. L., trans., *Hiroshima and Nagasaki*, New York (Basic Books-Harper/Colophon) 1981, by the Committee for the Compilation of Materials on Damage Caused by the Atomic Bombs in Hiroshima and Nagasaki, originally published by the cities of Hiroshima and Nagasaki.

Jarrett, Keith, *Survivor's Suite*, ECM Records, 1978, cover note.

Jung, Carl Gustav, *The Undiscovered Self*, Boston, MA (Atlantic/Little, Brown) 1958.

Jung, Carl Gustav, et al., *Man and His Symbols*, London (Aldus Books) 1964.

Jungk, Robert, *Brighter Than a Thousand Suns-A Personal History of the Atomic Scientists*, James Cleugh, trans., New York (Harvest/HBJ) 1958, originally published as *Heiler als Tausend Sonnen*, Bern (Alfred Scherz Verlag) 1956.

Katz, Michael, Marsh, William P., and Thompson, Gail Gordon, eds., *Earth's Answer-Explorations of Planetary Culture at the Lindisfarne Conferences*, New York (Harper & Row) 1977.

Kenner, Hugh, *The Pound Era*, Berkeley/Los Angeles (University of California Press) 1971.

Keyes, Ken, Jr., *The Hundredth Monkey*, Coos Bay, OR (Vision Books) 1982.

Keys, Donald, *Earth at Omega-The Passage to Planetization*, New York (Branden) 1985.
To Arrest Nuclear War, Menlo Park, CA (Planetary Citizens) 1982.

Keys, Donald, & Laszlo, R., *Disarmament, The Human Factor*, New York (Pergamon Press) 1981.

Koestler, Arthur, *Janus-A Summing Up*, New York (Vintage/Random House) 1978.

Kübler-Ross, Elizabeth et al., *Death-The Final Stage of Growth*, Englewood Cliffs, NJ (Prentice-Hall/Spectrum) 1975.

Kuhn, Thomas S., *The Structure of Scientific Revolutions*, Chicago/London (University of Chicago Press) 1962, 1970.

Lauer, Eugene, and Mlecko, Joel, Eds., *A Christian Understanding of the Human Person*, Ramsey, NJ (Paulist) 1982.

Leakey, Richard, *Origins*, New York (Dutton) 1977.

Long, Mary, "Visions of a New Faith," *Science Digest*, Nov. 1981.

Lorenz, Konrad, *L'Instinct dans le Comportment de l'Animal et de l'Homme*, Paris (Masson et Cie) 1956.

Lovejoy, Arthur O., *The Great Chain of Being*, Cambridge, MA (Harvard University Press) 1936, 1964.

Lovelock, James, *GAIA-A New Look at Life on Earth*, London/New York, 1972.

Luijpen, H., *Existential Phenomenology*, The Hague (Martinus Nijhof) 1972.

MacPherson, Malcolm C., *Time Bomb-Fermi, Heisenberg, and the Race for the Atomic Bomb*, New York (Dutton) 1986.

Mahoney, R., "The Catholic Conscience and Nuclear War," in *Commonweal*, CIX, 12 March 1982, pp. 138ff.

Malthus, Thomas R., *An Essay on the Principle of Population* (7th edition, 1872), New York (A. M. Kelley) 1971.

Marcel, Gabriel, *Homo Viator*, Paris (Aubier) 1944.

Maritain, Jacques, *Man and the State*, Chicago (University of Chicago Press) 1951.

Mayer, Milton S., *They Thought They Were Free-The Germans, 1933-45*, Chicago (University of Chicago Press) 1955.

McGinn, Bernard, *Apocalyptic Spirituality*, Ramsey, NJ (Paulist) 1979.

Milgrim, Stanley, *Obedience to Authority*, New York (Harper & Row) 1974.

Mische, Gerald and Patricia, *Toward a Human World Order-Beyond the National Security Straitjacket*, Ramsey, NJ (Paulist) 1977.

Mumford, Lewis, *The Myth of the Machine*, vol. 1: *Technics & Human Development*; vol. 2: *The Pentagon of Power*, New York (Harcourt, Brace, Jovanovich) 1967/70.
My Works and Days, New York (Harcourt, Brace, Jovanovich) 1979.

Needleman, Jacob, *Sin and Scientism*, Interview by Robert Briggs, San Francisco (Broadside Editions/Robert Briggs Associates) 1985.

Nishitani, Keiji, *Religion and Nothingness*, Jan Van Bragt, trans., Berkeley/Los Angeles (University of California Press) 1982.

Oppenheimer, J. Robert, "Physics and the Contemporary World," in *The Open Mind*, New York (Simon & Schuster) 1955.

Owens, Virginia Stem, *And the Trees Clap Their Hands*, Grand Rapids, MI (Eerdmans) 1983.

Panikkar, R., *El Silencio del Dios*, Madrid (Guadiana) 1970.
  The Trinity and the Religious Experience of Man, New York/London
  (Orbis/Darton, Longman & Todd) 1973.
  Worship and Secular Man, Maryknoll, NY (Orbis) 1973.
  The Vedic Experience, Berkeley/Los Angeles (University of California
  Press) 1977.
  The Intrareligious Dialogue, Ramsey, NJ (Paulist) 1978.
  Myth, Faith and Hermeneutics, Ramsey, NJ (Paulist) 1979.
  Blessed Simplicity, The Monk As Universal Archetype, Scott Thomas
  Eastham, ed., New York (Seabury) 1982.
  "The Existential Phenomenology of Truth," *Philosophisches Jahrbuch
  der Görres-Gesellschaft*, 64 Jahrgang 1956, pp. 27-54.
  "Die Zukunft kommt nicht später," in *Vom Sinn der Tradition*, L.
  Reinisch, ed., München (C. H. Beck) 1970, pp. 53-64.
  "Verstehen als Ueberzeugtsein," *Neue Anthropologie*, Hans-Georg
  Gadamer and Paul Vogler, eds., Vol. 7: *Philosophische
  Anthropologie*, Stuttgart (Georg Thieme Verlag) 1974, pp. 132-167.
  "Have Religions the Monopoly on *Religion?*," *Journal of Ecumenical
  Studies*, vol. XI, no. 3, Summer 1974, pp. 515-7.
  "Singularity and Individuality. The Double Principle of Individuation,"
  in *Revue Internationale de Philosophie*, 111-112, 1975, fasc. 1-2, pp. 141-166.
  "Man as a Ritual Being," in *Chicago Studies*, vol. XVI, no. 1, 1976, pp. 5-28.
  "Philosophy as Life-Style," in *Philosophes critiques d'eux-mêmes*, A.
  Mercier, ed., Bern (Peter Lang for FISP) 1978.
  "Colligite Fragmenta: For An Integration of Reality," in *From Alienation
  to At-Oneness*, F. Eigo, ed., Proceedings of the Theology Institute of
  Villanova University, New York (Villanova University Press) 1978,
  pp. 19-91.
  "Rtatattva: A Preface to a Hindu-Christian Theology," in *Jeevadhara*,
  no. 49, Jan.-Feb. 1979, pp. 6-63.
  "The Myth of Pluralism: The Tower of Babel-A Meditation on Non-
  violence," *Cross Currents*, vol. XXIX, no. 2, Summer 1979, pp. 197-230.
  "The End of History-Three Kairological Moments in Human Time-
  Consciousness," *Teilhard and the Unity of Knowledge*, Thomas F. King
  and James F. Salmon, eds., New York (Paulist) 1983, pp. 83-141.

Plato, *The Collected Dialogues*, Hamilton, Edith, and Cairns, Huntington,
  eds., Princeton (Bollingen/Pantheon) 1961.

Pound, Ezra, *The Cantos*, New York (New Directions) 1970.

Rolt, C. E., trans., *Dionysus the Areopagite*, London (SPCK) 1920.

Russett, B. M., & Blair, B. G., eds., *Progress in Arms Control?*, New York (W. H. Freeman) 1979.

Ruyter, Knut W., "Pacifism and Military Service in the Early Church," *Cross Currents*, vol. XXXII, no. 1, pp. 54-70.

Ryan, Tim, and Jappinen, Rae, eds., *The Whole Again Resource Guide*, annual periodical published by Source Net, Santa Barbara, CA (distributed by Capra Press) 1982.

Sampson, R. V., *The Psychology of Power*, New York (Vintage) 1968.

Saxon, David, ed., *The Effects of Nuclear War*, U.S. Congress, Office of Technology Assessment, Washington, D.C. 20510.

Scheler, Max, *Formalism in Ethics and Non-formal Ethics of Values*, Evanston, IL (Northwestern University Press) 1973.

Schell, Jonathan, *The Fate of the Earth*, New York (Knopf) 1982.

Shannon, T. A., ed., *War or Peace? The Search for New Answers*, Maryknoll, NY (Orbis) 1980.

Shibata, Shingo, *Alice Herz als Denkerin un Friedenskämpferin*, Amsterdam (Grüner) 1977.
"The Tasks of Philosophy and Social Science to Prevent Nuclear Extinction-Philosophy and Social Science for Survival," in his book *Tasks of Our Time*, vol. 1: *For Abolition of Nuclear Weaponry*, in Japanese, Tokyo, 1978. Partial English translation, "Religion & Nuclear Extermination," in *The Churchman*, Aug.-Sept. 1979, pp. 6-7.

Skeat, Walter W., *A Concise Etymological Dictionary of the English Language*, New York (Capricorn) 1963.

Smith, Huston, *Beyond the Post-Modern Mind*, New York (Crossroad) 1982.

Smuts, Jan Christian, *Holism and Evolution*, New York (Macmillan) 1926.

Steiner, George, *After Babel-Aspects of Language and Translation*, London (Oxford University Press) 1975.

Stevenson, W. Taylor, *History as Myth*, New York (Seabury) 1965.

Strachey, J., ed., *Standard Edition of the Complete Psychological Works of Sigmund Freud*, vol. 22, *Why War?*, London (Hogarth) 1933.

Streng, Frederick J., *Emptiness. A Study in Religious Meaning*, Nashville/New York (Abingdon) 1967.

Swimme, Brian, *The Universe Is a Green Dragon-A Cosmic Creation Story*, Santa Fe, NM (Bear & Co.) 1985.

Szent-Györgyi, Albert, *The Crazy Ape*, New York (Grosset & Dunlap) 1970.

Thomas, Lewis, *Lives of a Cell*, New York (Viking) 1974.

Tirman, J., ed., *The Fallacy of Star Wars: Why Space Weapons Can't Protect Us*, New York (Vintage) 1984.

Touchman, Barbara, *A Distant Mirror*, New York (Random House) 1978.

Tucker, R. W., *Just War and Vatican Council II: A Critique*, New York (The Council on Religion and International Affairs) 1967.

United Nations, *Comprehensive Study on Nuclear Weapons*, 1981 (Study Series 1), Department of Political and Security Council Affairs, UN Center for Disarmament, Report of the Secretary General, UN Publication sales no. E.81.I.11.
Disarmament, A Periodic Review by the United Nations, published quarterly by the Department for Disarmament Affairs, UN Sales Section, Room DC2-853, United Nations, New York, NY 10017.

Vonnegut, Kurt, *Slaughterhouse Five*, New York (Delacorte Press) 1969.

Waskow, Arthur, "Mystery Not Mastery," in *Co-Evolution Quarterly*, no. 39, Fall 1983, pp. 43-6.

Watson, Lyall, *Lifetide*, New York (Bantam) 1980.

Weil, Simone, *Waiting for God*, New York (G. P. Putnam's Sons) 1951.

Wilber, Ken, *The Holographic Paradigm and Other Paradoxes*, Boulder/Boston (Shambhala) 1982.
Quantum Questions-Mystical Writings of the Great Physicists, Boulder/Boston (Shambhala) 1982.

Williams, William Carlos, *Spring and All* (1923), Frontier Press, 1970.

Zukav, Gary, *The Dancing Wu Li Masters*, New York (Morrow) 1979.

About the Author

Born in Chicago, 1949, with a Ph.D. in Religious Studies from the University of California, Scott Thomas Eastham has for the past four years taught Religion & Culture at the Catholic University of America in Washington, D.C., and is currently Visiting Professor of Intercultural Communications at Western Maryland College, Westminster, MD. He is the author of *Paradise & Ezra Pound-The Poet as Shaman; Wisdom of the Fool*; and other works.

R. Panikkar

One of today's most acclaimed scholars of comparative religion, Raimundo Panikkar virtually embodies the inter-religious dialogue. He is equally at home in the Catholic, Hindu, and Buddhist worlds, and holds doctorates in the sciences, humanities, and theology. He is the author of some 30 books, including *The Unknown Christ of Hinduism; The Intrareligious Dialogue; Myth, Faith & Hermeneutics*; and *The Vedic Experience.*